Privacy in Social Networks

Synthesis Lectures on Data Mining and Knowledge Discovery

Editors

Jiawei Han, *University of Illinois at Urbana-Champaign*
Lise Getoor, *University of Maryland*
Wei Wang, *University of North Carolina, Chapel Hill*
Johanness Gehrke, *Cornell University*
Robert Grossman, *University of Chicago*

Synthesis Lectures on Data Mining and Knowledge Discovery is edited by Jiawei Han, Lise Getoor, Wei Wang, and Johannes Gehrke. The series publishes 50- to 150-page publications on topics pertaining to data mining, web mining, text mining, and knowledge discovery, including tutorials and case studies. The scope will largely follow the purview of premier computer science conferences, such as KDD. Potential topics include, but not limited to, data mining algorithms, innovative data mining applications, data mining systems, mining text, web and semi-structured data, high performance and parallel/distributed data mining, data mining standards, data mining and knowledge discovery framework and process, data mining foundations, mining data streams and sensor data, mining multi-media data, mining social networks and graph data, mining spatial and temporal data, pre-processing and post-processing in data mining, robust and scalable statistical methods, security, privacy, and adversarial data mining, visual data mining, visual analytics, and data visualization.

Giovanni Seni and John F. Elder 2010

Modeling and Data Mining in Blogosphere
Nitin Agarwal and Huan Liu 2009

Privacy in Social Networks

Elena Zheleva, Evimaria Terzi, and Lise Getoor

ISBN: 978-3-031-00773-6   paperback
ISBN: 978-3-031-01901-2   ebook

DOI 10.1007/978-3-031-01901-2

A Publication in the Springer series
SYNTHESIS LECTURES ON DATA MINING AND KNOWLEDGE DISCOVERY

Lecture #4

Series Editors: Jiawei Han, *University of Illinois at Urbana-Champaign*
                Lise Getoor, *University of Maryland*
                Wei Wang, *University of North Carolina, Chapel Hill*
                Johanness Gehrke, *Cornell University*
                Robert Grossman, *University of Chicago*

Series ISSN
Synthesis Lectures on Data Mining and Knowledge Discovery
Print 2151-0067   Electronic 2151-0075

Privacy in Social Networks

Elena Zheleva
LivingSocial

Evimaria Terzi
Boston University

Lise Getoor
University of Maryland, College Park

SYNTHESIS LECTURES ON DATA MINING AND KNOWLEDGE DISCOVERY #4

ABSTRACT

This synthesis lecture provides a survey of work on privacy in online social networks (OSNs). This work encompasses concerns of users as well as service providers and third parties. Our goal is to approach such concerns from a computer-science perspective, and building upon existing work on privacy, security, statistical modeling and databases to provide an overview of the technical and algorithmic issues related to privacy in OSNs. We start our survey by introducing a simple OSN data model and describe common statistical-inference techniques that can be used to infer potentially sensitive information. Next, we describe some privacy definitions and privacy mechanisms for data publishing. Finally, we describe a set of recent techniques for modeling, evaluating, and managing individual users' privacy risk within the context of OSNs.

KEYWORDS

privacy, social networks, affiliation networks, personalization, protection mechanisms, anonymization, privacy risk

Contents

Authors' Biographies

Acknowledgments

The authors would like to thank Michael Hay and Ashwin Machanavajjhala for the invaluable and thorough feedback on this manuscript. We would also like to thank the LINQS group at the University of Maryland, College Park and the data-management group at Boston University. This manuscript was supported in part by NSF grant #IIS0746930 and NSF grant #1017529, and gifts from Microsoft, Yahoo!, and Google. Some of the images included in Figures 3.1 and 5.1 were taken from Wikimedia Commons and FreeDigitalPhotos.net. The cartoon characters in Figure 2.3 and Figure 6.1 are from www.lumaxart.com. Finally, Figure 8.1 was taken from [66].

Elena Zheleva, Evimaria Terzi, and Lise Getoor
February 2012

CHAPTER 1

Introduction

Social-networking sites and other online collaborative tools have emerged as immensely popular places for people to post and share information. Facebook, Google+, MySpace, LinkedIn, etc., all have benefits, ranging from practical (e.g., sharing a business document via Google Docs) to purely social (e.g., communicating with distant friends via Facebook). At the same time, not surprisingly, information sharing poses real threats to user privacy. For example, in social-networking sites, documented threats include *identity theft*, *digital stalking*, and *personalized spam* [21]. College admissions officers have even withdrawn student admission offers using information posted on social networking sites [87]. Thus, online presence in social networks involves a trade-off between the benefits of sharing information with colleagues, friends, and acquaintances, and the risks that personal information is used in unintended, and potentially harmful, ways.

In the last few years, with the myriads of social media and social network websites appearing online, there has been a renewed and growing interest in understanding social phenomena rising from people's interactions and affiliations [3, 36]. These websites have thousands, and even millions, of users which voluntarily submit personal information in order to benefit from the services offered, such as maintaining friendships, blogging, sharing photos, music, articles, and so on. This rich information can also be used in a variety of ways to study people's personal preferences, patterns of communication and flow of information. Apart from facilitating sociological studies, the publicly available user information can be used to train predictive models, which can infer hidden information about individuals and possibly predict users' behavior. These models are widely used to improve the user experience within the social-networking websites. For example, a model that predicts what a Facebook user considers important can be used to judiciously select the pieces of information shown on the user's feed.

Although such models can be utilized towards the development of better personalization strategies, they inevitably raise privacy concerns. On the one hand, users want to have the best possible experience within the online social-networking sites. This means that strong user models need to be built; strong models also require information about the user: demographic information, behavioral information, information about the user's social context, etc. On the other hand, access to this information poses serious privacy concerns. While many users are happy to benefit from the aggregate information collected from millions of users for improving search results and recommendations, many are less comfortable with sharing their own information for the purpose of, say, targeted advertising. And, when it comes to making use of personal information for unintended means (to sell to third parties, as the basis for denying access to resources, etc.), there are even greater concerns.

This synthesis lecture addresses privacy concerns in the context of online social networks (OSNs). In this setting, there are several major categories of players including: users, service providers, and third parties interested in making use of the data. Each of these players can have benevolent or malevolent intentions. Benevolent users are typically interested in sharing and communicating information through online media. Malevolent users include scammers, stalkers, and identity thieves. Online media service providers have a variety of motives; they may be interested in mining data to provide additional utility to their users, or extracting information in order to produce goods they can sell to third party providers. Third parties are individuals or companies who are interested in user data for the purpose of advertising, market research, or collecting and re-selling the data. Again, this can be for benevolent or malevolent uses. For many uses, such as targeted advertising, the characterization depends on the perspective of the player—while many users are be put off by targeting advertising, others may consider it a useful attentional filter.

Our goal in this book is to provide a survey of work on privacy in OSNs that encompasses concerns of both users, service providers, and third parties. It approaches the problem from a computer science perspective, building on existing work in privacy, security, statistical modeling and databases to provide an overview of the technical issues relating to privacy in OSNs. We present a framework for understanding information disclosure in OSNs, catalog of privacy breaches, methods for privacy-preserving publication, models for user information sharing.

In Part I of this book, we begin by providing background which describes OSNs, the sensitive user information that can be potentially (inadvertently) disclosed, and common methods for inferring information from network data. We begin by introducing a simple data model for OSNs (Chapter 2), introduce a framework for describing information disclosure in OSNs (Chapter 3), and review some of the common statistical inference methods in networks, as well as show how they relate to the different forms of information disclosure (Chapter 4). These methods can be used in both a benevolent manner to infer missing information in order to improve services, or they can be viewed as attacks by an adversary.

The underlying motivation for the privacy definitions and publishing mechanisms, present in Part II, is from the point of view of the service providers who are interested in providing data access to third parties while maintaining user anonymity. Privacy research has a long history, and one of the major challenges has been in defining privacy. In Part II, we give an overview of privacy definitions (Chapter 5) and privacy-preserving techniques for publishing social network data in Chapter 6.

The focus of Part III shifts from the service providers and third parties to the privacy-management concerns of users. In this part of the survey, we give an overview of methods for modeling, evaluating and managing users' privacy risk. Chapter 7 describes a model for information sharing which captures the trade-off between the benefits of sharing and the risks of unwanted information propagation. Chapter 8 discusses methods for evaluating users' privacy risk, and Chapter 9 discusses the management of users' online privacy settings. Existing work on these aspects of privacy is currently somewhat limited and, therefore, each chapter within Part III focuses and summarizes on a specific piece of work on user-specific privacy management.

Addressing the privacy challenges for online social networks is an active new area of research [55]. However, privacy research has a longer history in the data mining, database, and security communities that we are not covering in this synthesis lecture. For example, privacy-preserving data mining aims at creating data mining algorithms and systems which take into consideration the sensitive content of the data [4, 97]. Chen et al. [24] provide a comprehensive, recent survey of the field of privacy-preserving data publishing. The database and security communities have studied interactive and non-interactive mechanisms for sharing potentially sensitive data [32]. Most of this research assumes that there are one or more data owners who would like to provide data access to third parties while meeting privacy constraints. In contrast, access to data in OSNs is often freely available, and users can specify their personal privacy preferences.

There are many important issues related to privacy in OSNs that we are also not covering. For example, we are not touching upon issues related to policy making. That is, we do not discuss the types of policies that can be put in place by governments and other agencies, so that the privacy of individuals is protected by law. Neither are we discussing philosophical issues related to the contradictory needs of people to protect their privacy and also share information with their peers, or ethical issues, related to the collection and analysis of large online social network data [30]. Finally, we also do not discuss surveillance techniques that (potentially illicitly) monitor users' behavior online and in social networks [41]. These issues are beyond the scope of this survey. Here, we focus more narrowly on the modeling and the algorithmic aspects of privacy and social networks.

In addition to the work presented in this synthesis lecture, there are other useful surveys on privacy in social networks that focus on different aspects of the problem [27, 49, 64, 101, 108]. The surveys on privacy-preserving data publication for networks [64, 101, 108] cover privacy attacks, edge modification, randomization and generalization privacy-preserving strategies for network structure. Clarkson et al. [27] discuss anonymization techniques which aim to prevent identity disclosure. The survey of Hay et al. [49] concentrates on privacy issues with network structure, and it covers attacks and their effectiveness, anonymization strategies, and differential privacy for private query answering. Another recent summary of existing work, that also includes users' privacy risk studies, is presented in the tutorial by Hay et al. [48]. Further, researchers from psychology, communication, sociology, and information science provide perspectives on the balance between people's need to communicate and bond with others while creating private spaces online in a recent book [96].

Our work differs from the above in that it is the first that combines in one place a discussion of privacy issues in social media from the perspective of modeling, protecting and managing. This information is scattered across the literature in data mining, databases, theoretical computer science and statistics. We hope the survey provides a concise introduction to this spectrum of topics that will be of use to users, practitioners, and providers.

PART I

Online Social Networks and
Information Disclosure

Privacy in online social networks is a very young field that studies what constitutes an unauthorized disclosure in social networks and how to protect sensitive user information. In this part, we provide the necessary background for the synthesis lecture. We begin by presenting a simplified social network model in Chapter 2. Then, we discuss the sensitive information that can be potentially (inadvertently) disclosed in Chapter 3. We introduce a framework for describing information disclosure in online social networks by formally defining four types of privacy breaches: *identity disclosure, attribute disclosure, social link disclosure*, and *affiliation link disclosure*.

While inferring personal user information may have potential privacy implications, it is also a centerpiece in many personalized online services and ad targeting techniques. Therefore, discovering hidden knowledge in social networks is a compelling application of machine learning and link mining algorithms. In Chapter 4, we review some of the common statistical inference tasks in networks and show how these relate to the different forms of information disclosure. These methods can be used in both a benevolent manner to infer missing information in order to improve services, or they can be viewed as attacks by an adversary.

CHAPTER 2

A Model for Online Social Networks

In the context of this book, when we refer to social networks, we generally mean online social networks. This includes online sites such as Facebook, Twitter, LinkedIn, Google+, and Flickr, where individuals can create user accounts and link to, or "friend," each other, and which allow rich interactions such as joining communities or groups of interest, or participating in discussion forums. These sites often include services which allow users to build online profiles and share their preferences and opinions about items, such as tagging articles and postings, commenting on photos, and rating movies, books, or music.

Figure 2.1 shows a hypothetical user profile on the online social network Facebook. On these profiles, you can often see many personal details about the users. For example, Emily's profile displays her gender, where she lives, when she was born, and her religious and political views. On her profile, there is also a list of people she has befriended on Facebook. The online social network is defined by these declared friendship links' with other people. In addition to the personal attributes and online friendships, user profiles on social networks often show other online content that people add and comment on, such as photos, movies, fan pages, etc. Often, services also provide the ability to form user-defined groups, and users have the option of becoming members or affiliating with these groups.

These different types of entities can be described in an entity-relationship (ER) model, which the database community uses for representing data at an abstract level. Figure 2.2 shows an ER diagram of an online social network with six types of entities. Entities are represented as boxes, and relationships are represented as lines between them. The ER diagram represents both the online social network of a user (users, profiles, online groups, and pages), and the real-world social network of an individual (individuals and groups). It highlights the distinction between the online and real-world networks, and we want to emphasize that there is not necessarily one-to-one mapping between the two. We focus on the online portion of the social network, above the dashed line in Figure 2.2, along with the mapping between the online social users and the individuals in the real world.

The set of real-world individuals P is represented by the entity type *Individual*. An individual p_i is an instance of this type, and it can be described by a number of personal attributes, denoted by $p_i. A$, such as name, age, gender, and political affiliation. Individuals can create user accounts in online social networks, and V corresponds to the set of entities of type *User*. Each *User* entity, v_i, has a *Profile*, which contains attributes $v_i.Profile.A$ declared by the account holder. We use the shorthand notation $v_i.A$ for $v_i.Profile.A$. For every user v_i and every profile item a the user has a *Privacy Setting* (or simply *Setting*) denoted by $R(v_i.a)$. The setting encodes the set of people that v_i is willing to share information about attribute $v_i.a$. For example, $R(v_i.a)$ can be set to private or public, or visible by friends, friends-of-friends, etc. We also use $R(v_i.A)$ to refer to the *Profile Privacy Settings*. H is the set of entities of type *Group* that represents affiliation groups. Each group h_k has a group *Page*. Groups have attributes $h_k.Page.A$, denoted by $h_k.A$, such as a name, category and formation date. Users can link to other users in the network, and the relation *is a friend with* represents these links as user-user pairs $e_v(v_i, v_j) \in E_v$. A user can be affiliated with multiple groups in the online social network, and each group has to have at least one user. The relation *is affiliated with* represents this affiliation as user-group pairs $e_h(v_i, h_k) \in E_h$. The ER model can also include other types of entities, such as articles, movies, and brands, that users affiliate with, which we omit here for simplicity. However, some of the work that we review in this book includes these types of affiliations. Similarly, we allow for other types of online user-user relationships, such as colleagues, family relationships, and co-authorships, even thought we have not shown them in the ER diagram.

Figure 2.1: A hypothetical Facebook profile.

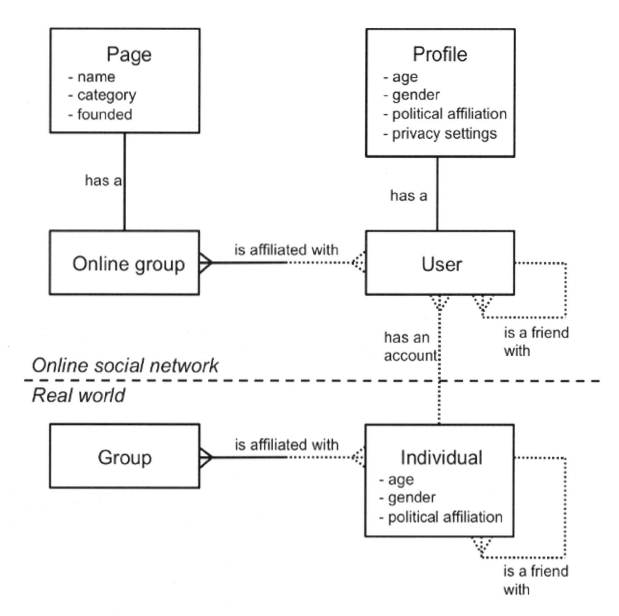

Figure 2.2: An entity-relationship (ER) diagram of social and affiliation networks which highlights the distinction between the online and real-world networks. Dashed lines signify whether a relationship of this type is optional for the entity it connects, and a solid line shows that the relationship is mandatory for the entity. Different shapes at the end of the lines represent the cardinality of the relationship.

As a running example, we consider the social network presented in Figure 2.3. It consists of seven profiles which describe a collection of users (Ana, Bob, Chris, Don, Emma, Fabio, and Gina), along with their social links and their affiliations with online groups. Users are linked by a friendship link, and in this example they are reciprocal. There are two online groups that users participate in: the "Espresso lovers" affiliation group and the "Yucatan" affiliation group. These individuals also have personal attributes on their profiles, such as name, age, gender, zip code, and political views (we have not shown attributes in Figure 2.3; see Figure 5.2 on p. 27 for an example). User-group affiliations can also be represented as a bipartite graph, such as the ones in Figure 6.1 (p. 38) and Figure 6.2(a) (p. 39).

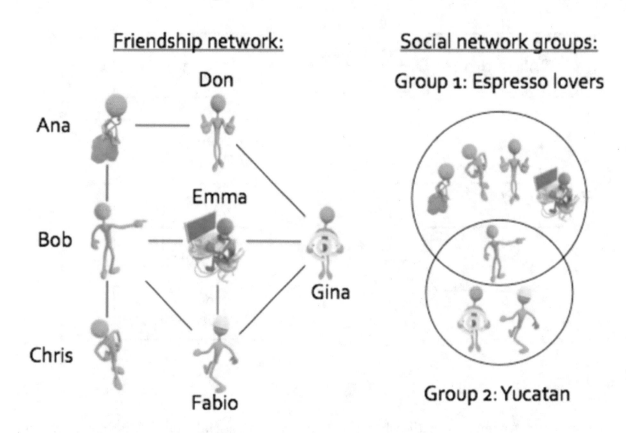

Figure 2.3: A toy social and affiliation network.

In the coming chapters, we will be discussing privacy in social networks in the context of the ER model we have just described.

CHAPTER 3

Types of Privacy Disclosure

When studying privacy, it is important to define a failure to preserve privacy. A *privacy breach* occurs when a piece of sensitive information about an individual is disclosed to an adversary, someone whose goal is to access information that they are not authorized to access. Traditionally, two types of privacy breaches have been studied: *identity disclosure* and *attribute disclosure*. We discuss these two types in the context of social networks. We also present two additional disclosure types, specific to network data: *social link disclosure* and *affiliation link disclosure*.

3.1 IDENTITY DISCLOSURE

Identity disclosure occurs when an adversary is able to determine the mapping from a user v in the social network to a specific real-world entity p. Before we are able to provide a formal definition of identity disclosure, let us consider three questions related to the identity of p in which an adversary may be interested.

Definition 3.1 Mapping query. For a particular individual p, find which user $v \in V$ in a social network maps to p. Return v.

Definition 3.2 Existence query. For a particular individual p, find if this individual corresponds to a user v in the network. Return *true* or *false*.

Definition 3.3 Co-reference resolution query. For two users v_i and v_j, find if they refer to the same individual p. Return *true* or *false*.

A simple way of defining *identity disclosure* is to say that the adversary can answer the *mapping query* correctly and with full certainty. However, unless the adversary knows unique attributes of individual p that can be matched with the observed attributes of users in V, this is hard to achieve. One way of formalizing *identity disclosure* for an individual p is to associate a random variable $\hat{v}_p$ which ranges over the users in the network. We assume that the adversary has a way of computing the probability of each user v_i referring to individual p, $Pr(\hat{v}_p = v_i)$. In addition, we introduce a dummy user v_{dummy} in the network which serves the purpose of absorbing the probability of individual p not having a profile in the network. We assume that p has exactly one user account, and the true user account of p in $V \cup \{v_{dummy}\}$ is v_*. Even though we de not discuss the case of multiple user accounts per individual, the definitions can be extended to handle this case. We use the shorthand $Pr_p(v_i) = Pr(\hat{v}_p = v_i)$ to denote the probability that v_i corresponds to p; Pr_p provides a mapping $Pr_p : V \cup \{v_{dummy}\} \rightarrow \mathbb{R}$. We leave it open as to how the adversary constructs Pr_p. Then we can define *identity disclosure* as follows.

Definition 3.4 Identity disclosure with confidence t. In a set of individual profiles V in a social network, identity disclosure occurs with confidence t when $Pr_p(v_*) \geq t$ and $v_* \neq v_{dummy}$.

An alternative definition of *identity disclosure* considers that the possible values of v_p can be ranked according to their probabilities.

Definition 3.5 Identity disclosure with *top-k* confidence. In a set of users V in a social network, identity disclosure occurs with *top-k* confidence when v_* appears in the top k users (or top $p\% = k * 100/|V|$), in the list of users ranked by Pr_p from high to low.

The majority of research in social network privacy has concentrated on identity disclosure [8, 17, 22, 50, 51, 57, 65, 79, 99, 102, 107, 109]. We discuss it in more detail in Chapter 6.

3.2 ATTRIBUTE DISCLOSURE

A common assumption in the privacy literature is that there are three types of possibly overlapping sets of personal attributes:

- Identifying attributes—attributes, such as social security number (SSN), which identify an individual uniquely.

- Quasi-identifying attributes—a combination of attributes which can identify an individual uniquely, such as name and address.

- Sensitive attributes—attributes that individuals may like to keep hidden from the public, such as political affiliation and sexual orientation.

Usually, it is assumed that real-world, true attributes of an individual, $p_i.A$ are the same as the observed attributes of a user profile corresponding to this individual, $u_i.A$. While an attribute value of a user profile may correspond to the true individual attribute value, the user profile may also have missing and incorrect attribute values, e.g., an individual may declare an incorrect date of birth or location on his user profile. Here, we would like to point this distinction but for ease of exposition, we will assume that they are the same.

Attribute disclosure occurs when an adversary is able to determine the true value of a sensitive user attribute, one that the user intended to stay private. Friendship links, groups, and user-group affiliations can also have sensitive attributes, which we do not cover here. However, the definitions we discuss next are easily generalizable. To make the attribute disclosure definition more concrete, we assume that each sensitive user attribute $v_i.a_s$ for user v has an associated random variable $v_i.\hat{a}_s$ which ranges over the possible values for $v_i.a_s$. The true value of $v.a_s$ is $v.a_*$. We also assume that the adversary can map the set of possible sensitive attribute values to probabilities, $Pr_a(v.\hat{a}_s = v.a) : \rightarrow \mathbb{R}$, for each possible value $v.a$. Note that this mapping can be different for each user. Now, we can define attribute disclosure as follows.

Definition 3.6 Attribute disclosure with confidence t. For a user v with a hidden attribute value $v.a_s = v.a_*$, attribute disclosure occurs with confidence $Pr_a(v.\hat{a}_s = v.a_*) \geq t$.

Similarly to *identity disclosure*, there is an alternative definition of *attribute disclosure* which considers that the possible values of $v.A_s$ can be ranked according to their probabilities.

Definition 3.7 Attribute disclosure with *top-k* confidence. For a user v with a hidden attribute value $v.a_s = v.a_*$, attribute disclosure occurs with *top-k* confidence when a_* appears in the top k values of the list of possible values ranked by their probabilities Pr_a.

Clearly, if an adversary can see the identifying attributes in a social network, then answering the identity *mapping query* becomes trivial, and identity disclosure with confidence 1 can occur. For example, if a user profile contains a SSN, then identifying the individual behind the profile is trivial since there is a one-to-one mapping between individuals and their social security numbers. Therefore, in order to prevent identity disclosure, the identifying attributes have to be removed from the profiles.

Sometimes, a combination of the quasi-identifying attributes can lead to identity disclosure. What constitutes quasi-identifying attributes depends on the context. For example, it has been observed that 87% of individuals in the U.S. Census from 1990 can be uniquely identified based on their date of birth, gender and zip code [94]. Another example of quasi-identifiers is a combination of a person's name and address.

Similarly, matching records from different datasets based on quasi-identifying attributes can lead to further privacy breaches. This is known as a *linking attack*. If the identities of users in one dataset are known and the second dataset does not have the identities but it contains sensitive attributes, then the sensitive attributes of the users from the first dataset can be revealed. For example, matching health insurance records, in which the identifying information is removed, with public voter registration records can reveal sensitive health information about voters. Using this attack,

Sweeney matched the two types of records using zip code, birth date and gender, and thus, she was able to identify the medical record of the governor of Massachusetts [94].

In the context of social and affiliation networks, there has not been much work on sensitive attribute disclosure. Most studies look at how attributes can be predicted [62, 78, 106], and very few on how they can be protected [22]. We discuss this work in more detail in Chapter 6.

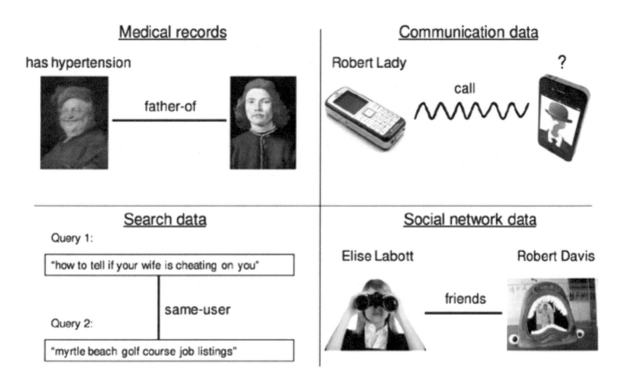

Figure 3.1: Sensitive link examples.

3.3 SOCIAL LINK DISCLOSURE

Social link disclosure occurs when an adversary is able to find out about the existence of a sensitive relationship between two users, a relationship that these users would like to remain hidden from the public. Similarly to the previous types of disclosures, we assume that there is a random variable $\hat{e}_{i,j}$ associated with the link existence between two users v_i and v_j, and an adversary has a model for assigning a probability to $\hat{e}_{i,j}$, $Pr(\hat{e}_{i,j} = true) : e_{i,j} \rightarrow \mathbb{R}$.

Definition 3.8 Social link disclosure with confidence t. For two users v_i and v_j, a social link disclosure occurs with confidence t when $e_v(v_i, v_j) \in E_v$ and $Pr(\hat{e}_{i,j} = true) \geq t$.

Note that since the link existence $\hat{e}_{i,j}$ has only two possible values, true and false, the *top-k* definition does not apply to social link disclosure.

Examples of sensitive relationships can be found in online social networks, communication data, inheritable disease data and others. In social network data, some users prefer that their friendship links remain hidden from the public. In cell phone communication data, users may consider their cell phone calls to be of sensitive nature, e.g., finding that an unknown user has made phone calls to a cell phone number of another, known user can compromise the identity of the unknown user. In hereditary or sexually transmitted disease data, knowing the relationships between individuals who have been diagnosed with the diseases and ones that have not, can help infer the probability of the healthy individuals to develop these diseases. Figure 3.1 presents a summary of these examples.

Note that online friends are not the same as real-world friends, and there is a subtle difference between disclosing online social links and social links in the real world. This is an important issue for researchers to be cognizant of. However, the majority of the privacy literature does not make this distinction, and for the purposes of this synthesis lecture, we assume that online and real-world social links are the same.

Researchers have studied attacks that expose sensitive links in social networks [8, 13, 58, 105] that we discuss in Section 6. Sensitive edge properties, such as link strength (or weight), have also been the focus of recent work [31, 67].

3.4 AFFILIATION LINK DISCLOSURE

Another type of privacy breach in networks is *affiliation link disclosure*: whether a person belongs to a particular affiliation group. Whether two users are affiliated with the same group can also be of sensitive nature, e.g., when users belong to a prosecuted political group or a group centered around an unconventional user preference. Thus, hiding affiliations can be a key to preserving the privacy of individuals.

As before, we assume that there is a random variable $\hat{e}_{v,h}$ associated with the existence of an affiliation link between a user v and a group h, and that an adversary has a way of computing the probability of $\hat{e}_{v,h}$, $Pr(\hat{e}_{v,h} = true) : e_{v,h} \rightarrow \mathbb{R}$.

Definition 3.9 Affiliation link disclosure with confidence t. For a user v and an affiliation group h, an affiliation link disclosure occurs with confidence t when $e_h(v, h) \in E_h$ and $Pr(\hat{e}_{v,h} = true) \geq t$.

Sometimes, affiliation link disclosure can lead to another type of disclosure. For example, Wondracek et al. [99] show a de-identification attack in which affiliation link disclosure can lead to the identity disclosure of a supposedly anonymous Internet user. An adversary starts the attack by crawling a social networking website and collecting information about the online social group memberships of its users. It is assumed that the identities of the social network users are known. According to the collected data, each user who participates in at least one group has a group signature, which is the set of groups he belongs to. Then, the adversary applies a *history stealing attack* (for more details on the attack, see [99]) which collects the web browsing history of the target Internet user. By finding the group signatures of social network users which match the browsing history of the Internet user, the adversary is able to find a subset of potential social network users who may be the Internet user. In the last step of the attack, the adversary looks for a match between the id's of the potential users and the browsing history of the target individual, which can lead to de-identification of the Internet user.

Another example of affiliation link disclosure leading to identity disclosure is in search data. If we assume that users posing queries to a search engine are the individuals in the social network, and the search queries they pose are the affiliation groups, then disclosing the links between users and queries can help an adversary identify people in the network. Users interact with search engines in an uninhibited way and reveal a lot of personal information in the text of their queries. There was a scandal in 2006 when AOL, an Internet Service provider, released an "anonymized" sample of over half a million users and their queries posed to the AOL search engine. The release was well intentioned and meant to boost search ranking research by supplementing it with real-world data. Each user was specified by a unique identifier, and each query contained information about the user identifier, search query, the website the user clicked on, the ranking of that website in the search results, and the timestamp of the query. Here is a snapshot of the released data with omitted timestamps:

Table 3.1: A snapshot of the data released by AOL. Here, we are omitting the timestamps included in the data.

User ID	Search query	Clicked website	Ranking
4417749	clothes for age 60	http://www.news.cornell.edu	10
4417749	dog who urinate on everything	http://www.dogdayusa.com	6
4417749	landscapers in lilburn ga.		
4417749	pine straw in lilburn ga.	http://gwinnett-online.com	9
4417749	gwinnett county yellow pages	http://directory.respond.com	1
4417749	best retirement place in usa	http://www.amazon.com	7
4417749	mini strokes	http://www.ninds.nih.gov	1

One of the problems with the released data was that even though it was in a table format, its entries were not independent of each other. Shortly after the data release, New York Times reporters linked 454 search queries made by the same individual which gave away enough personal information to identify that individual—Thelma Arnold, a 62-year old widow from Lilburn, Georgia [10]. Her queries included names of people with the same last name as hers, information about retirement, her location, etc.

Affiliation link disclosure can also lead to attribute disclosure, as illustrated in a *guilt-by-association attack* [29]. This attack assumes that there are groups of users whose sensitive attribute values are the same, thus recovering the sensitive value of one user and the affiliation of another user to the group can help recover the sensitive value of the second user. This attack was used in the BitTorrent file-sharing network to discover the downloading habits of users [25]. Communities were detected based on social links, and monitoring only one user in each community was enough to infer the interests of the other people in the community. In this case the sensitive attribute that users would like to keep private is whether they violate copyrights. This attack has also been applied to identifying fraudulent callers in a phone network [29]. Cormode et al. [28] study data anonymization to prevent affiliation link disclosure. They refer to affiliation links as associations (see Section 6.2).

CHAPTER 4

Statistical Methods for Inferring Information in Networks

While inferring personal user information may have potential privacy implications, it is a centerpiece in many personalized online services and ad targeting techniques. Therefore, discovering hidden knowledge in social networks is a compelling application of machine learning and link mining algorithms, and it has been studied extensively in recent years. Link mining refers to data mining techniques that explicitly consider links between data entities when building predictive or descriptive models of the linked data [39, 104]. Commonly addressed link mining tasks include entity resolution, collective classification, link prediction and group detection. Namata et al. present a survey of link mining techniques for analyzing noisy and incomplete networks [77]. Next, we give a brief overview of these techniques and their relationship to the four types of privacy disclosures we discussed. The described methods can be used in both a benevolent manner to infer missing information in order to improve services, or they can be viewed as attacks by an adversary.

Machine learning and statistical inference methods are one set of techniques for inferring personal information. There are other approaches, such as adversarial attacks on private information in social networks, and we discuss them in more detail in Chapter 6.

4.1 ENTITY RESOLUTION

Entity resolution refers to the problem of reconciling data references corresponding to the same entity. Entity resolution is also known as the "correspondence problem" in the vision research community [81], the "co-reference resolution" problem in the natural language processing community [92], "deduplication" in the database research community [52], and "record linkage" in the statistics community [53].

Entity resolution on user profiles may lead to identity disclosure. For example, this happens when an individual (or a real-world entity) considers his online social network user profile (or a data reference) to be private information, known only to a selected few acquaintances online, but entity resolution gives an accurate answer to a mapping query (discussed in Section 3.1). Entity resolution can also identify user profiles of the same real-world individual by analyzing data from different social networks (similar to answering the co-reference resolution query in Section 3.1).

Entity resolution is an important first step in many link mining algorithms. Its goal is to reconstruct a clean network which removes duplicate entries and captures the relations among the true underlying entities in the graph [16]. The survey of Namata et al. [77] distinguishes between three categories of entity resolution approaches in network data: attribute-based, naive relational, and collective relational. Attribute-based approaches are the traditional approaches to entity resolution which rely solely on the attributes of the user profiles. Given two users, these approaches make use of similarity measures, such as a string similarity measure, or a weighted combination of multiple similarity measures, over the attributes of user pairs. Several sophisticated similarity measures have been proposed for entity resolution, and they use different types of features and domain knowledge. More recently, naive and collective relational approaches, which consider the links between users, have been proposed. The naive relational approaches consider the attribute similarity of connected users. The collective relational approaches, on the other hand, use the links to make decisions jointly. A major issue in entity resolution is that it is a known hard problem computationally if we want to consider all pairs of users. There have been a number of approaches proposed for blocking [52] and canopy approaches [74]. Namata et al. [77] give an overview of the research on performing entity resolution efficiently.

4.2 COLLECTIVE CLASSIFICATION

Supervised learning, or classification, refers to learning a machine learning model from labeled training data in order to infer the labels of unknown, testing data. When the predicted labels are sensitive personal attributes of users, classification can lead to attribute disclosure.

In the last decade, there has been a growing interest in supervised classification that relies not only on the attributes of a single entity but also on the attributes of the entities it is linked to, some of which may be unobserved [15, 90, 95]. Link-based or collective classification breaks the assumption that data comprises of independent and identically distributed (i.i.d.) entities. It can take advantage of autocorrelation and homophily in network data, the properties that make the attribute values of linked entities correlated with each other. Social networks are one of the domains where link-based classification can be applied because personal attributes and roles of connected people are often correlated. For example, political affiliations of friends tend to be similar, students tend to be friends with other students, etc. Comprehensive reviews of link-based classification can be found in the works by Sen et al. [90] and Bhagat et al. [15].

In a typical classification setting, there are data entities (or users) with two types of attributes: a label and a set of features. The label is an attribute value of the users that we are interested to predict, e.g., gender. Features are other, independent attributes of the users, e.g. weight, height, and hair length, which are given as an input to the classifier. The users with observed labels comprise the training data, and they can be used to train the classifier to predict the unobserved labels. What distinguishes link-based classification from traditional classification is that some of the users' features are based on the features or labels of neighboring users in the network.

Sen et al. [90] define collective classification as a combinatorial problem in which, given the class labels of some users in the network, the task is to infer the class labels of the rest of the users. They identify two types of approaches to collective classification. The first type relies on local conditional classifiers to perform approximate inference. Iterative classification algorithms [69, 80] and Gibbs sampling-based algorithms [72, 75] fall into this first category. Iterative classifiers rely on attributes of the users, some of which are observed and based on the labels of neighboring user, to infer an initial labeling of the unlabeled user and then use this labeling to update the user attribute values. This process continues iteratively until the assignment of labels stabilizes or the iterations reach a pre-specified threshold. Some models use a simplified version of Gibbs sampling which also relies on local classifiers and iterates through each user and estimates its probability distribution conditioned on its neighbor labels.

The second approach to collective classification is based on defining a global optimization function. One type of model which relies on a global optimization function is a pairwise Markov Random Field (MRF) [95], an undirected probabilistic graphical model. In particular, each user's label in the social network corresponds to a random variable in the MRF, and each user-user link is considered as a pairwise dependency between two random variables. Inference on the MRF can be used for classification of the missing attributes in the data.

Bhagat et al. [15] discuss a third type of models for user classification in social networks which fall into the category of information diffusion processes. These can be models which propagate user labels using random walks. The assumption in these models is that the probability of a user label is equal to the probability that a random walk starting from the user in question will end at another user in the network with that label.

4.3 LINK PREDICTION

Link prediction is the problem of predicting which pairs of users are likely to form social links. Link prediction can lead to privacy concerns when the predicted link is between users who consider this link to be private. In this case, a sensitive link disclosure occurs. Therefore, studying how to prevent sensitive link disclosure while providing accurate link recommendations is an important problem. Machanavajjhala et al. address this problem in their work [71], and we discuss it in more detail in Section 6.1.3.

Link prediction in social networks is useful for a variety of tasks. The most straightforward use is for making data entry easier—a link-prediction system can propose links, and users can select the friendship or professional links that they would like to include, rather than users having to enter the links manually. Link prediction is also a core component of any system for dynamic network modeling—the dynamic model can predict which users are likely to gain popularity, and which are likely to become central according to various social network metrics.

Lü and Zhou [68], as well as Al Hasan and Zaki [45], have written recent surveys on the topic of link prediction. Lü and Zhou distinguish between similarity-based algorithms, maximum likelihood methods, and approaches based on probabilistic models. Al Hasan and Zaki categorize approaches into feature-based link prediction, Bayesian probabilistic models, probabilistic relational models, and linear algebraic methods.

Usually, link-prediction algorithms process a set of features in order to learn and predict whether it is likely that two users in the data are linked. Sometimes, these features are hand-constructed by analyzing the problem domain, the attributes of the users, and the relational structure around those users [2, 44, 61, 84]. Other times, they are automatically generated, i.e., the prediction algorithm first learns the best features to use and then predicts new links [82].

When link prediction is posed as a pair-wise classification problem, one of the fundamental challenges is dealing with the large outcome space; if there are n users, there are n^2 possible social links. In addition, because most social networks are sparsely connected, the prior probability of any link is extremely small, thus we have to contend with a large class skew problem. Furthermore, constructing a representative training set is also challenging because the number of the negative instances is huge compared to the number of positive instances.

4.4 GROUP DETECTION

Another problem that often occurs in reasoning about network data is inferring the underlying hidden groups or community structures in the network. Group detection can be defined as the clustering of social network users into groups or communities [91]. When the group affiliations are of sensitive nature, as discussed in Section 3.4, group detection can lead to affiliation link disclosure.

Some group detection algorithms enforce that a node can be a member of only a single group, while others allow nodes to be members of more than one group. A community can be defined as a group of users that share dense connections among each other, while being less tightly connected to users in different communities in the network. The importance of communities lies in the fact that they can often be closely related to modular units in the system that have a common function, e.g., groups of individuals interacting with each other in a society, WWW pages related to similar topics, or proteins having the same biological function within the cell.

Namata et al. [77] distinguish between clique-finding techniques, clustering techniques, centrality-based techniques, and modularity-based techniques for group detection. As the name suggests, clique-finding methods rely on the discovery of subgraphs in the social network in which each user has a link to all other users in the subgraph. On the other hand, clustering techniques group users according to a specific similarity measure over their attributes and require that cluster possess high intra-cluster similarity and low inter-cluster similarity. Centrality-based techniques discover groups of users with denser connections by removing network links with high number of shortest paths running through them. Modularity-based techniques for group detection also discover groups of users with denser connections by maximizing the difference between the number of edges within the extracted communities against the expected number of edges in these communities in a random network.

Other sophisticated methods include using probabilistic graphical models [6], such as mixed-membership stochastic block models, which consider communities to be latent variables which have to be learned. The survey on statistical network models of Goldenberg et al. [40] includes an overview of such methods.

In this chapter, we discussed models for link mining in social networks. These models can be used for inference which benefits social network users, for example for content recommendation algorithms and personalized services. Because of the potentially sensitive nature of the data, these models can also be used for adversarial attacks. We discuss some of the adversarial scenarios in Chapter 6.

PART II

Data Publishing and
Privacy-Preserving Mechanisms

The underlying motivation for the privacy definitions and publishing mechanisms that we present in Part II are from the point of view of the service providers who are interested in providing access to their data while maintaining user anonymity. While privacy research has a long history, privacy preservation in the context of online social network data is a relatively new research field. Rather than assuming data which is described by a single table of independent user records with attribute information for each row, it takes into consideration more complex real-world datasets. As discussed earlier, network data, often represented as a collection of relational tables describing entities and their relations, implies richer dependencies between entities. Sanitizing such data requires understanding of the dependencies and removing sensitive information which can be inferred by direct or indirect means.

One of the major challenges in privacy research in general has been in defining privacy. In Chapter 5, we give an overview of existing privacy definitions for publishing social network data. In Chapter 6, we present privacy-preserving techniques for publishing social network data and existing adversarial attacks. We catalog existing work based on their assumptions for the data.

CHAPTER 5

Anonymity and Differential Privacy

The goal of data mining is discovering new and useful knowledge from data. Sometimes, the data contains sensitive information, and it needs to be sanitized before it is published publicly in order to address privacy concerns. Data sanitization is a complex problem in which hiding private information trades off with utility reduction. The goal of sanitization is to remove or perturb the attributes of the data which may help an adversary infer sensitive information, while maintaining the utility of the data for knowledge discovery. The solution depends on the properties of the data and the notions of privacy and utility in the data.

One way in which data providers can sanitize data is by anonymization. Figure 5.1 shows a typical scenario in which a data owner, such as a social network service provider, is interested in allowing third parties to access (parts of) their valuable data. In order to meet privacy concerns, she consults a privacy analyst before publishing a perturbed version of the data. In the process of anonymizing the data, the identifying information is removed and other attributes are perturbed. The privacy analyst needs to consider the original data representations, what the possible privacy breaches are, and what the desired post-anonymization data utility is. While this process looks relatively simple, it involves addressing an important trade-off: the goal of perturbing the data to prevent undesirable privacy breaches competes with the goal of preserving the useful knowledge in it.

Anonymizing techniques have been criticized as often being ad hoc and not providing a principled mechanism for preserving privacy. There are no guarantees that an adversary would not be able to come up with an attack which uses background information and properties of the data, such as node attributes and observed links, to infer the private information of users. Another way of sanitizing data is by providing a private mechanism for accessing the data, such as allowing algorithms which are provably privacy-preserving to run on it. Next, we will discuss privacy preservation definitions. Some of these definitions were not developed specifically for network data but we provide examples from the social network domain.

To formalize privacy preservation, Chawla et al. [23] propose a framework based on the intuitive definition that "our privacy is protected to the extent we blend in the crowd." Obviously, with the richness of information in online social network profiles, this is hard to achieve and users are often easily identifiable. We will look at a simpler case when a data provider is interested in releasing a dataset with online social network profiles. To give a flavor of existing work, we present four existing privacy definitions which make the notion of "blending in the crowd" more precise.

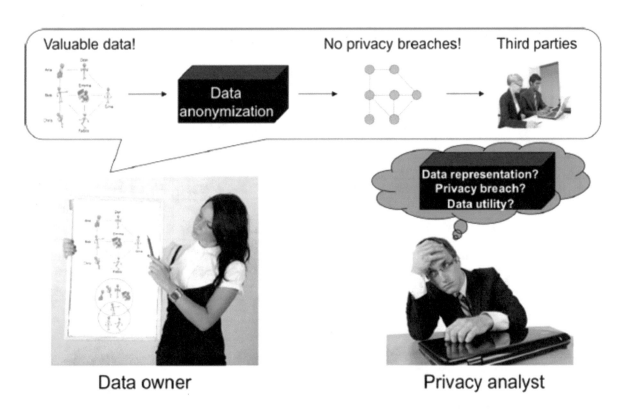

Figure 5.1: Anonymization scenario in which a data owner consults a privacy analyst before providing third parties with a perturbed version of her valuable data.

5.1 *k*-ANONYMITY

k-anonymity protection of data is met if the information for each individual contained in the data cannot be distinguished from at least $k - 1$ other individuals in the data [88, 89, 94]. *k*-anonymity can be achieved by suppressing and generalizing the attributes of users in the data. Suppressing an attribute value means deleting it from the perturbed data and replacing it with a wildcard value that matches any possible attribute value. Generalizing an attribute means replacing it with a less specific but semantically consistent value. One can see suppression as a special case of generalization, and that suppressing all attributes would guarantee *k*-anonymity. This is why a notion of utility in the data has to be incorporated whenever sanitizing data. The actual objective is to maximize utility by minimizing the amount of generalization and suppression. Achieving *k*-anonymity by generalization with this objective as a constraint is an NP-hard problem [5]. *k*-anonymity has been studied mostly for table data, so we begin by presenting its definition using only the nodes *V* and their attributes *V.A*, i.e., disregarding links and affiliation groups.

Identifier	Quasi-identifiers			Sensitive
Name	Age	Sex	Zip	Pol. views
Ana	21	F	20740	liberal
Bob	25	M	83222	liberal
Chris	24	M	20742	liberal
Don	29	M	83209	conservative
Emma	24	F	20640	liberal
Fabio	24	M	20760	liberal
Gina	28	F	83230	liberal
Halle	29	F	83201	conservative
Ian	31	M	83220	conservative
John	24	M	20740	liberal

5-anonymity applied to data →

Equiv. class	Quasi-identifiers			Sensitive
	Age	Sex	Zip	Pol. views
C1	[21,24]	*	20***	liberal
C2	[25,31]	*	832**	liberal
C1	[21,24]	*	20***	liberal
C2	[25,31]	*	832**	conservative
C1	[21,24]	*	20***	liberal
C1	[21,24]	*	20***	liberal
C2	[25,31]	*	832**	liberal
C2	[25,31]	*	832**	conservative
C2	[25,31]	*	832**	conservative
C1	[21,24]	*	20***	liberal

Figure 5.2: 5-anonymity applied to data with 10 records.

Definition 5.1 *k*-anonymity. A set of records *V* satisfies k-anonymity if for every tuple $v \in V$ there exist at least $k - 1$ other tuples $v_{i_1}, v_{i_2}, \ldots, v_{i_{k-1}} \in V$ such that $v_{i_1}.A_q = v_{i_2}.A_q = \ldots = v_{i_{k-1}}.A_q$ where $A_q \in A$ are the quasi-identifying attributes of the profile.

Figure 5.2 shows an example of applying 5-anonymity to the data describing 10 individuals in the simple case where there are no social links between them. The data includes their names, ages, genders, and zip codes. The perturbed data meets a 5-anonymity constraint because each individual is indistinguishable from at least 4 other individuals. Here, the assumption is that name is an identifying attribute, therefore it has been suppressed. Three of the attributes, *Age*, *Sex*, and *Zip code*, are quasi-identifiers, therefore, they have been generalized. The sensitive attributes remain the same.

k-anonymity provides a clustering of the nodes into equivalence classes such that each node is indistinguishable in its quasi-identifying attributes from some minimum number of other nodes. In the previous example, there were two equivalence classes: class *C*1 of individuals whose age is in the range [21, 24] years and have a zip code 20 * **, and class *C*2 of individuals whose age is in the range [25, 31] years and have a zip code 832 * *. Note, however, that these equivalent classes are based on node attributes only, and inside each equivalence class, there may be nodes with different identifying structural properties and edges. This makes it hard to define k-anonymity for nodes in social networks. We discuss some approaches later in Chapter 6.

k-anonymity ensures that individuals cannot be uniquely identified by a linking attack. However, it does not necessarily prevent sensitive attribute disclosure. Here, we present two possible attacks on *k*-anonymized data [70]. The first one can occur when there is little diversity in the sensitive attributes inside an equivalence class. In this case, the sensitive attribute of everyone in the equivalence class becomes known with high certainty. For example, if an adversary wants to figure out Ana's political views knowing that her age is 21 and her zip code is 20740, then he can figure out that her record is in equivalence class *C*1. There is no diversity in the sensitive attribute value of equivalence class *C*1, i.e., everyone in *C*1 has liberal political views, therefore, the adversary is able to infer Ana's political views even though he does not know which row corresponds to her. This is known as the *homogeneity attack* [70].

The second problem with *k*-anonymity is that in the presence of background knowledge, attribute and identity disclosure can still occur. For example, knowing that someone's friends are liberal, makes it highly likely that this person is liberal as well. In our toy example, the knowledge that Gina's friends, Emma and Fabio, belong to equivalence class C1 where everyone is liberal, can help an adversary infer with high certainty that Gina is liberal as well. This is known as the *background attack* [70].

There are a number of definitions derived from *k*-anonymity tailored to structural properties of network data. Some examples include *k-degree anonymity* [65], *K-Candidate anonymity* [51], *k-automorphism anonymity* [109], *k-neighborhood anonymity* [101, 107], and *(k,l)-grouping* [28]. We introduce the intuition behind them, together with their definitions in Section 6.1.1 and Section 6.2, privacy mechanisms for networks.

5.2 *l*-DIVERSITY AND *t*-CLOSENESS

A privacy definition which alleviates the problem of sensitive attribute disclosure inherent to k-anonymity is *l*-diversity [70]. As its name suggests, *l*-diversity ensures that the sensitive attribute values in each equivalence class are diverse.

Definition 5.2 *l*-diversity. A set of records in an equivalence class C is *l*-diverse if it contains at least *l* "well-represented" values for each sensitive attribute. A set of nodes V satisfy *l*-diversity if every equivalence class $C' \subseteq V$ is *l*-diverse.

There are a number of ways to define "well-represented." Some examples include using frequency counts and measuring entropy. However, even in the case of *l*-diverse data, it is possible to infer sensitive attributes when the sensitive distribution in a class is very different from the overall distribution for the same attribute. If the overall distribution is skewed, then the belief of someone's value may change drastically in the anonymized data (*skewness attack*) [60]. For example, only 30% of the records in Figure 5.2 have conservative political views. However, in equivalence class C2 this number becomes 60%, thus the belief that a user is conservative increases for users in C2. Another possible attack, known as the *similarity attack*

[60], works by looking at equivalent classes which contain very similar sensitive attribute values. For example, if *Age* is a sensitive attribute and an adversary wants to figure out Ana's age knowing that she is in equivalence class *C*1 (based on her *Zip code*), then he would learn that she is between 21 and 24 years old which is a much tighter age range than the range in the whole dataset.

This leads to another privacy definition, *t*-closeness, which considers the sensitive attribute distribution in each class, and its distance to the overall attribute distribution [60]. The distance can be measured with any similarity score for distributions.

Definition 5.3 *t*-closeness. A set of records in an equivalence class C is *t*-close if the distance between the distribution of a sensitive attribute A_s in C and its distribution in V is no more than a threshold t. A set of nodes V satisfy *t*-closeness if every equivalence class $C' \subseteq V$ is *t*-close.

Just like with *k*-anonymity, sanitizing data to meet either *l*-diversity or *t*-closeness comes with a computational complexity burden. There are other privacy definitions of this flavor but they have all been criticized for being ad hoc. While they guarantee syntactic properties of the released data, they come with no privacy semantics [33].

5.3 DIFFERENTIAL PRIVACY

The notion of differential privacy was developed as a principled way of defining privacy, so that "the risk to one's privacy [. . .] should not substantially increase as a result of participating in a database" [32]. This shifts the view on privacy from comparing the prior and posterior beliefs about individuals before and after publishing a database to evaluating the risk incurred by joining a database. It also imposes a guarantee on the data release mechanism rather than on the data itself. Here, the goal is to provide statistical information about the data while preserving the privacy of users in the data. This privacy definition gives guarantees that are independent of the background information and the computational power of the adversary.

To illustrate differential privacy, first we assume a simpler case in which the data contains of social network users with their attributes only (*V.A*) and no links between them. At the end of this section, we also discuss recent work which goes beyond this assumption and looks at differential privacy for complex social network data. If the data is released using a differentially private mechanism, this would guarantees that Ana's participation in the social network does not pose a threat to her privacy because the statistics would not look very different without her participation. However, it *does not* guarantee that one cannot learn sensitive information about Ana using background information [32].

Definition 5.4 ϵ-differential privacy. A randomized function K satisfies ϵ-differential privacy if for all data sets D_1 and D_2 differing in at most one element, and any subset S of possible outcomes in *Range(K)*,

$$P(K(D_1) \in S) \le exp(\epsilon) \times P(K(D_2) \in S). \tag{5.1}$$

Here, a user profile is an element, thus $D_1 \subseteq V$ and $D_2 \subseteq V$. The randomized function K can be thought of as an algorithm which returns the value of a random variable, possibly with some noise. When developing a differentially private algorithm, one has to keep in mind the utility of the data and incorporate the desired knowledge in the algorithm. *Range(K)* is the output range of algorithm K. A common way of achieving ϵ-differential privacy is by adding random noise to the query answer.

One type of algorithm that has been proven to be differentially private is a *count* query to which one adds Laplacian noise [35]. For example, if the count query is $K =$ *"How many people are younger than 22?"*, then the output range of the query is *Range(K)* $= \{1, \ldots, n\}$ where n is the size of the social network. The count query is considered a low-sensitivity query because it has a sensitivity of $\Delta K = 1$ for any D_1 and D_2 differing in one element. Sensitivity is defined as

$$\Delta K = \max_{D_1, D_2} \|K(D_1) - K(D_2)\| \tag{5.2}$$

for any D_1 and D_2 which differ in at most one element. Note that this query has the same sensitivity not only for our specific data but for any data in this format. The Laplacian noise, which is added to the answer, is related to the sensitivity of the query.

A *mean* query, such as K = *"What is the average age of people in the social network?"*, has an even lower sensitivity for large data sets because removing any profile from the social network would change the output of the query by at most ΔK = *max(age)/n*. There are also queries, such as *median* queries, which have high sensitivity and require different techniques for generating noise.

A similar and somewhat weaker definition of differential privacy is the one of (ϵ, δ)-differential privacy which was developed to deal with very unlikely outputs of K [34].

Definition 5.5 (ϵ,δ)-differential privacy. A randomized function K satisfies ϵ-differential privacy if for all data sets D_1 and D_2 differing in at most one element, and any subset S of possible outcomes in *Range(K)*,

$$P(K(D_1) \in S) \leq exp(\epsilon) \times P(K(D_2) \in S) + \delta. \qquad (5.3)$$

Generally, ϵ and δ are considered to be very small numbers and are picked according to different considerations, such as the size of the database.

Recent work analyzes the challenges of applying differential privacy to social networks include links between entities. For example, Kifer and Machanavajjhala [54] argue that it is not possible to guarantee privacy and utility without making assumptions about the data. They distinguish between user participation in the data generation and the actual presence of a user in the data, and show that "privacy of an individual is preserved when it is possible to limit the inference of an attacker about the participation of the individual in the data generating process."

We illustrate this distinction with our toy social network (Figure 2.3), where D_1 includes all seven users and their links, while D_2 excludes Bob and his links. It may still be possible to infer Bob's presence by analyzing other people's links. If an adversary knows that Bob introduced Emma to

Fabio, then Bob caused their link generation, and thus, his presence can be inferred.

On the other hand, Hay et al. [47] investigate whether network statistics, such as degree and clustering coefficient, can be measured reliably under differential privacy guarantees. They introduce a number of variants of differential privacy definitions specifically adapted to graph data. One of them, k-edge ϵ-differential privacy, prevents an adversary from learning about edges in the graph. Hay et al. also introduce a differentially private algorithm which allows provably accurate estimates of the degree sequence of a graph [46].

5.4 OPEN PROBLEMS

Publishing complex network data with privacy guarantees remains a challenge. For example, adapting differential privacy to network data is not straightforward [46, 47, 54]. The development of (possibly new) *rigorous* privacy definitions which address the complexity of network data is a very important research direction.

CHAPTER 6

Attacks and Privacy-preserving Mechanisms

So far, we have discussed existing privacy definitions. Next, we present existing privacy mechanisms for social and affiliation networks. Except for the privacy mechanisms based on differential privacy, each mechanism was developed to counteract specific adversarial attacks and background knowledge. We organized this chapter based on the data assumptions that authors make. Section 6.1 covers social networks in which there are users and social links (V and E_v), and Section 6.2 includes an overview of the mechanisms for affiliation networks (V, H, and E_h). Finally, we describe research which considers both social and affiliation links in Section 6.3.

6.1 PRIVACY MECHANISMS FOR SOCIAL NETWORKS

The majority of research on privacy mechanisms for social networks considers anonymization which strips off all the attributes from user profiles but keeps some of the structure based on the social links between users [8, 50, 51, 65, 107, 109]. We describe this research in Section 6.1.1. In Section 6.1.2, we mention approaches to anonymizing data which consider that there is utility in keeping both user attributes and information about social links [22, 105].

6.1.1 ANONYMIZING NETWORK STRUCTURE

One naïve way of anonymizing a social network is by removing all the attributes of the user profiles, and leaving only the social link structure. This creates an anonymized graph which is isomorphic to the original graph. The intuition behind this approach is that if there are no identifying profile attributes, then attribute and identity disclosures cannot occur, and thus the privacy of users is preserved. Contrary to the intuition, this not only removes a lot of important information but it also does not guarantee the privacy of users. Two types of attacks have been proposed which show that identity and social link disclosures can occur when it is possible to identify a subgraph in the released graph in which all the node identities are known [8]. An *active attack* assumes that an adversary can insert accounts in the network before the data release, and a *passive attack* assumes that a number of friends can collude and share their linking patterns after the data release.

In the active attack an adversary creates k accounts and links them randomly, then he creates a particular pattern of links to a set of m other users that he is interested to monitor. The goal is to learn whether any two of the monitored nodes have links between them. When the data is released, the adversary can efficiently identify the subgraph of nodes corresponding to his k accounts with provably high probability. Then he can recover the identity of the monitored m nodes and the links between them which leads to social link disclosure for all $\binom{m}{2}$ pairs of nodes. With as few as $k = \Theta(\log n)$ accounts, an adversary can recover the links between as many as $m = \Theta(\log^2 n)$ nodes in an arbitrary graph of size n. The passive attack works in a similar manner. It assumes that the exact time point of the released data snapshot is known and that there are k colluding users who have a record of what their links were at that time point.

Another type of structural background information that can be exploited is similar in spirit to the linking attack mentioned in Section 3.2. The existence of an auxiliary social network in which the identity of users is known can help an adversary identify nodes in a target social network [79]. Starting from a set of users which form a clique both in the target and the auxiliary networks, an adversary expands the matching by finding the most likely nodes that correspond to each other in the two networks by using structural information, such as number of user friends (node degree), and number of common neighbors. To validate this attack, it has been shown that the discovered matches sometimes correspond to matches found using descriptive user attributes such as username and location in the social networks of Twitter and Flickr [79].

Structural privacy. Starting from the idea that certain subgraphs in the social network are unique, researchers have studied the mechanism of protecting individuals from identity disclosure when an adversary has *background information about the graph structure* around a node of interest [50, 51, 65, 102, 107, 109]. Each node has structural properties (subgraph signature) that are the same as the ones of a small set of other nodes in the graph, called a candidate set for this node [51]. Knowing the true structural properties of a node, an adversary may be able to discover the identity of that node in the anonymized network. Structure queries can be posed to the network to discover nodes with specific subgraph signatures.

Looking at the immediate one-hop neighbors, each node has a star-shaped subgraph in which the size of the subgraph is equal to the degree of the node plus one. With the assumption that identity disclosure can occur based on a node's degree, the degree becomes an identifying attribute that a data provider would like to hide. In our toy network (Figure 2.3), Ana and Don would be in each other's candidate sets because they both have degree 2; Emma, Gina, and Fabio appear in the same candidate set for either of the three nodes; Bob and Chris are uniquely identifiable because they are the only ones in the network with degrees four and one, respectively. The notion of *k-degree anonymity* [65] was formulated to protect individuals from an adversary who has background information of user's node degrees. It states that each node should have the same degree as at least $k - 1$ other nodes in the anonymized network.

Adding the links between the one-hop neighbors of a node, sometimes referred to as the 1.5-hop neighborhood, creates a richer structural signature. Based on this, Ana and Don still have the same subgraph signature, and so do Emma and Fabio. However, Gina has a unique signature and is easily identifiable by an adversary who has knowledge of her true 1.5-hop neighborhood structure. Zhou and Pei [107] formalize the desired property to protect individuals from this type of attack. A graph satisfies *k-neighborhood anonymity* if every node in the network has a 1.5-hop neighborhood graph isomorphic to the 1.5-hop neighborhood graph of at least $k - 1$ other nodes. The name of this property was given by Wu et al. [101].

In our example, Ana and Don become uniquely identifiable once we look at their 2-hop neighborhoods. Emma and Fabio have isomorphic signatures regardless of the size of the neighborhood for which the adversary has background information. This leads to the most general privacy preservation definitions of *K-Candidate anonymity* [51] and k-automorphism anonymity [109].

Definition 6.1 K-Candidate anonymity. An anonymized graph satisfies K-Candidate Anonymity with respect to a structural query Q if there is a set of at least K nodes which match Q, and the likelihood of every candidate for a node in this set with respect to Q is less than or equal to $1/K$.

Structural queries are subgraph signatures that can identify users. Hay et al. define three types of structural queries: vertex refinement queries, subgraph queries and hub fingerprint queries [50, 51]. Zou et al. [109] assume a much more powerful adversary who has knowledge of any subgraph signature of a target individual. They propose the notion of *k-automorphism anonymity* to fend off such an adversary.

Definition 6.2 k-automorphism anonymity. An anonymized graph is k-automorphic if every node in the graph has the same subgraph signature (of arbitrary size) as at least $k - 1$ other graph nodes, and the likelihood of every candidate for that node is less than or equal to $1/k$.

Anonymization. The anonymization strategies for social network structure fall into four main categories.

- **Edge modification**. Since complete removal of the links to keep structural properties private would yield a disconnected graph, edge modification techniques propose edge addition and deletion to meet desired constraints. Liu and Terzi anonymize the network degree sequence to meet k-degree anonymity [65]. This is easy to achieve for low-degree nodes because the degree distribution in social networks often follows a power law. For each distinguishable higher-degree node, where distinguishable is defined as a degree for which there are less than k nodes with that degree, the anonymization algorithm increases its degree artificially so that it becomes indistinguishable from at least $k - 1$ other nodes. The objective function of the algorithm is to minimize the number of edge additions and deletions. We discuss another edge modification algorithm [107] with a similar objective but a stronger privacy guarantee in Section 6.1.2. Zou et al. [109] propose an edge modification algorithm that achieves *k*-automorphism anonymity.

- **Randomization**. Anonymization by randomization can be seen as a special case of anonymization by edge modification. It refers to a mechanism which alters the graph structure by removing and adding edges at random, and preserves the total number of edges. Hay et al. [51] show that if this is performed uniformly at random, then it fails to keep important graph metrics of real-world networks. Ying and Wu [102] propose *spectrum-preserving randomization* to address this loss of utility. The graph's spectral properties are the set of eigenvalues of the graph's adjacency matrix to which important graph properties are related. Preserving this spectrum guides the choice of random edges to be added and deleted. However, the impact of this approach on privacy is unclear.

  Two recent studies have presented algorithms for reconstructing randomized networks [98, 100]. The goal of reconstruction is given an observed, randomized network, to find the most probable true graph from which the randomized network was created. Wu et al. [100] take a low-rank approximation approach and apply it to a randomized network structure, such that accurate topological features can be recovered. They show that in practice reconstruction may not pose a larger threat to privacy than randomization because the original network is more similar to the randomized network than to the most probable, reconstructed network. Vuokko and Terzi [98] consider reconstruction mechanisms for networks where randomization has been applied both to the structure and attributes of the nodes. They identify cases in which reconstruction can be achieved in polynomial time. Both reconstruction strategies can be considered as attacks on network anonymization by randomization.

- **Network generalization**. One way to alleviate an attack based on structural background information is by publishing the aggregate information about the structural properties of the nodes [50]. In particular, one can partition the nodes and keep the density information inside and between parts of the partition. Nodes in each partition have the same structural properties, so that an adversary coming with a background knowledge is not able to distinguish between these nodes. In practice, sampling from the anonymized network model creates networks which keep many of the structural properties of the original network, such as degree distribution, path length distribution and transitivity. Network generalization strategies for other network types are discussed in Section 6.1.2 [22, 105] and Section 6.3 [13].

- **Differentially private mechanisms**. Differentially private mechanisms refer to algorithms which guarantee that individuals are protected under the definition of differential privacy (see Section 5.3). Hay et al. [47] propose an efficient algorithm which allows the public release of one of the most commonly studied network properties, degree distribution, while guaranteeing differential privacy. The algorithm involves a post-processing step on the differentially private output, which ensures a more accurate result. The empirical analysis on real-world and synthetic networks shows that the resulting degree-distribution estimate exhibits low bias and variance, and can be used for accurate analysis of power-law distributions, commonly occurring in networks.

6.1.2 ANONYMIZING USER ATTRIBUTES AND NETWORK STRUCTURE

So far, we have discussed anonymization techniques which perturb the structure of the network but do not consider attributes of the nodes, such as gender, age, nationality, etc. However, providing the (perturbed) structure of social networks is often not sufficient for the purposes of the researchers who study them. In another line of privacy research, the assumption is that anonymized data will have utility only if it contains both structural properties and node attributes.

Anonymization. Zhou and Pei [107] assume that each node has one attribute which they call a label. They show that achieving k-neighborhood *generalization* algorithm. The algorithm extracts the 1.5-neighborhood signatures for all nodes in the graph and represents them concisely using *DFS trees*. Then it clusters the signatures and anonymizes the ones in each cluster to achieve *k*-neighborhood anonymity. The objective function of the algorithm is similar to the one of Liu and Terzi [65], the minimization of the number of edge additions.

Zheleva and Getoor [105] studied the problem of social link disclosure in graphs with multiplex relations. The assumption is that an adversary has an accurate statistical model for predicting sensitive relationships if given the attributes of nodes and edges in the original data, therefore attributes have to be perturbed in the released data. They propose anonymization by generalization of the data as a two-step process. In the first step, nodes are treated as a table of records, and their attributes are anonymized to guarantee the privacy of users, for example, to meet one of the privacy definitions described earlier. Using *k*-anonymity, this creates a partition of the network nodes into equivalence classes. In the second step, the structure of the network is partially preserved by keeping aggregate structural information inside and between the equivalence classes.

Campan and Truta [22] also take a network generalization approach to anonymizing a social network. Their greedy algorithm optimizes a utility function using the attribute and structural information simultaneously rather than as a two-step process. They introduce a structural information loss measure, and adopt an existing measure of attribute information loss. The anonymization algorithm can be adjusted to preserve more of the structural information of the network or the nodes' attribute values.

6.1.3 PRIVACY OF SOCIAL RECOMMENDATION ALGORITHMS

To enhance the online experience of users, social media providers often offer personalized services, such as recommendations and tailored content. In social networks, these services can rely on friends' preferences in order to gauge the personal preferences of users, e.g., a movie that a friend watched or a blog that a friend read may be recommended. Since this can reveal personal information that these friends did not intend to reveal, personalized services based on social network information bring a lot of privacy concerns.

Machanavajjhala et al. [71] address the problem of recommending links, e.g., online friendships, while protecting information about other, sensitive links which are not visible to the two users to whom the recommendation is made. They use the definition of differential privacy and provide lower bounds on the minimum loss in utility. They show that in order to protect privacy, accurate social recommendations are possible only for a small subset of social network users. While high-degree nodes are most likely to receive accurate recommendations under the discussed private recommendation mechanisms, most nodes in social networks have relatively low node degrees which follow a power-law distribution, and they cannot benefit from these mechanisms.

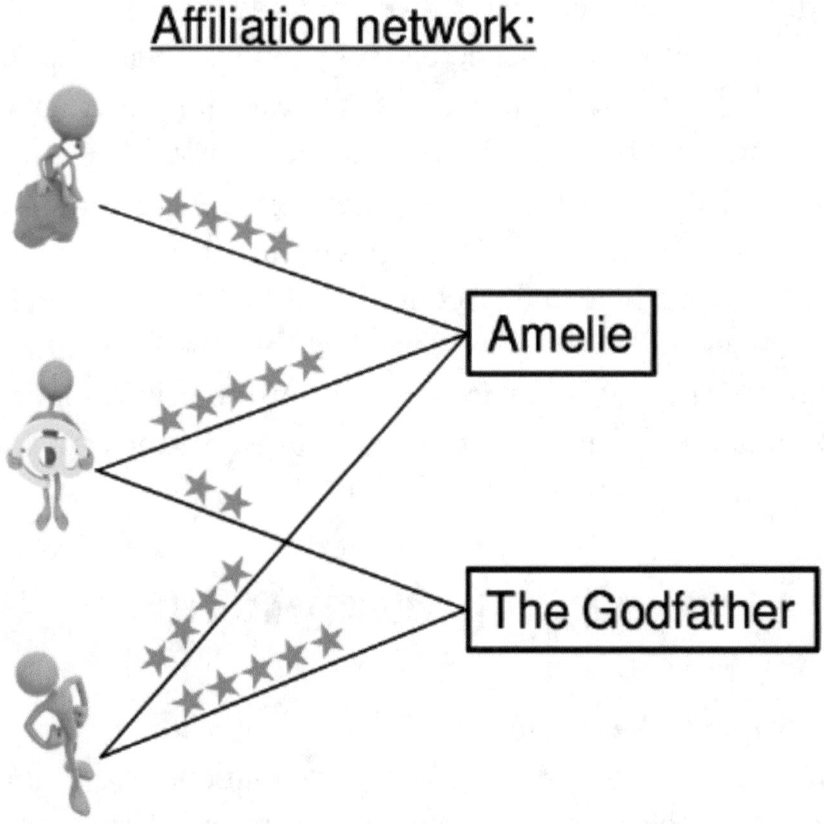

Figure 6.1: An affiliation network as a bipartite graph between three users and two movies. The affiliation links show the ratings that users gave to the movies on a scale from 1 to 5.

6.2 PRIVACY MECHANISMS FOR AFFILIATION NETWORKS

Next, we concentrate on the case when the data represents an affiliation network (users V, groups H and affiliation links E_h), and discuss privacy-preserving techniques developed specifically for this type of network. The affiliation network can be represented as a bipartite graph with two types of nodes V and H, and the affiliation links between them E_h. Figure 6.1 shows one example for this type of graph with users and movies that the users rated. The affiliation links have a weight corresponding to the movie rating for each user, on a scale from 1 to 5.

A well-known example for such a network is a dataset released by Netflix, an online movie rental company. Netflix set up a competition aimed at improving their movie recommendation systems. They released a dataset with around 100 million dated ratings from 480,000 randomly chosen Netflix customers. To protect customer privacy, each customer id was replaced with a randomly-assigned id. Unfortunately, this naive anonymization was found to be vulnerable under a linking attack [78]. Using the dates of user ratings and matching the records released by Netflix and user profiles in IMDB, an online movie database, Narayanan and Shmatikov [78] were able to achieve identity and sensitive attribute disclosure for some of the users in the Netflix dataset.

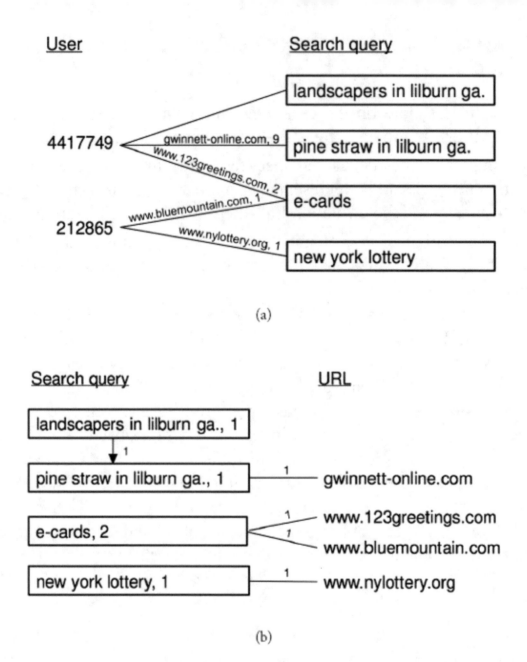

Figure 6.2: (a) *User-query graph* representing the users, their queries, the websites they clicked on and the ranking of each website; and (b) its reformulation into a *search query graph*.

A related problem is the problem of releasing a search query graph in which user information is contained in the affiliation links between search engine queries and clicked website URLs [57]. In particular, there is a bipartite graph of (query, URL) pairs. Here, the links have weights corresponding to the number of users who posed a particular query and clicked on the particular URL. In addition, there are links between queries with a weight equal to the number of users who posed the first query and then reformulated it into the second query. Each query also has counts of the number of times the query was posed to the search engine. The utility in such data is in using it for learning better search ranking algorithms. Figure 6.2(a) shows an example a user-query graph. Figure 6.2(b) shows its reformulation into a search query graph where individual users are not represented explicitly but only as aggregate numbers.

6.2.1 ANONYMIZATION

Two types of privacy mechanisms for affiliation networks have been studied in the research literature.

- **Network generalization**. Cormode et al. [28] propose a privacy definition for affiliation networks, *(k,l)-grouping*, tailored to prevent sensitive affiliation link disclosure. The authors make the assumption that affiliation links can be predicted based on node attributes and the structure of the network. They show why existing table anonymization techniques fail to preserve the structural properties of the network, and propose a greedy anonymization algorithm which keeps the structure intact but generalizes node attributes. The algorithm requires that each node is indistinguishable from at least $k - 1$ other nodes in terms of node properties, and each affiliation group is indistinguishable from at least $l - 1$ other affiliation groups, the basis of (k, l)-grouping. The utility is in being able to answer accurately aggregate queries about users and affiliation groups.

- **Differentially private mechanisms**. A private mechanism for a recommender system has been developed specifically for the movie recommendation setting [76]. The system works by providing differentially private mechanisms for computing counts, rating averages per movie and per user, and the movie-movie covariances in the data. These statistics are sufficient for computing distances based on k-nearest neighbor for predicting the ratings associated with new affiliation links. Using the statistics released by the mechanism, the algorithm performs with an accuracy comparable to the one in the original data.

Korolova et al. [57] propose an (ϵ,δ)-differentially private algorithm which allows the publication of a search query graph for this purpose. Here, the search logs are the database, and pairs of databases D_1 and D_2 are considered to differ in one element when one database excludes the search logs of exactly one user. The algorithm keeps only a limited number of queries and clicks for each user and allows for two types of functions on the graph which are sufficient for evaluating ranking algorithms. The first function gives a search query and its noisy count if it exceeds a pre-specified threshold. The second function publishes the noisy weight of the (query, URL) link for the top URLs for each query which was safe to publish according to the first function.

6.3 PRIVACY MECHANISMS FOR COMPLEX NETWORKS

There has not been much research on the privacy implications of the interplay between social and affiliation networks. It is obvious that they inherit all the privacy issues discussed so far for either type of network. What is not so obvious is that the complex dependencies these networks create can allow an adversary to learn private information in intricate ways. In particular, one can use the social environment of users to learn private information about them. One type of attack, which we call an *attribute inference attack*, assumes that an attribute is sensitive only for a subset of the users in the network and that the other users in the network are willing to publish it publicly [106]. The analogy in real-world social networks is the existence of private and public user profiles. The attack works by creating a statistical model for predicting the sensitive attribute using the publicly available information, and applying that model to predict the users with private profiles. In its basic form, the attack assumes that besides the network structure, the only user attributes that are available are the sensitive attribute value for the public profiles. Naturally, using other profile attributes can create even more powerful statistical models, as Lindamood et al. show [62]. An adversary succeeds when he can recover the sensitive

By taking into account all social and affiliation links, often declared publicly in online social networks, the model can use link-based classification techniques. As discussed in Section 4.2, link-based classification breaks the assumption that data comprises of independent and identically distributed (iid) user data points, and it can take advantage of autocorrelation, the property that attributes of linked objects often correlated with each other. For example, political affiliations of friends tend to be similar, students tend to be friends with other students, etc. Zheleva and Getoor [106] applied such techniques to predict user attributes, such as gender, location, and political views, in datasets from four popular online social networks, Facebook, Flickr, Dogster, and Bibsonomy. Their results suggest that link-based classification can predict sensitive attributes with high accuracy using information about online social groups, and that social groups have a higher potential for leaking personal information than social links.

Anonymization. Bhagat et al. [13] consider attacks for sensitive social link disclosure in social and affiliation networks, to which they refer as *rich interaction graphs*. Two nodes participating in the same group is also considered as a sensitive social link between the two users. Bhagat et al. represent the social and affiliation networks as a bipartite graph, in which one type of nodes are the users and the other type of nodes are affiliation groups. Social links are represented as affiliation groups of size two.

They propose two types of network generalization techniques to prevent social link disclosure. The first technique, a *uniform list approach*, keeps the structure intact, in a manner similar to (k, l)-groupings [28]. It divides nodes into classes of size m ensuring that each node's interactions fall on nodes of different classes. Each class is split into label lists of size k, thus ensuring that the probability of a link between two users (through a social link or a common affiliation group) is at most $1/k$. If the adversary has a background knowledge of the identities of r of the nodes and k is equal to m, then this probability becomes $1/(k-r)$. The second technique, a *partitioning approach*, also divides the nodes into classes of size m so that each node's interactions fall on nodes of different classes. However, it does not keep the original network structure, and publishes only the number of edges between partitions. The probability of a link between two users is guaranteed to be at most $1/m$ with or without background knowledge. The utility of the anonymized graph is in allowing accurate structural and attribute analysis of the graph.

6.4 OPEN PROBLEMS

In the last few years, there has been a growing interest in studying privacy mechanisms for social networks. A number of attacks have been discovered, and a few solutions proposed. The majority of the work concentrates on developing variants of k-anonymity, and it often does not consider the data utility for valuable statistical analyses. Also, as discussed in Chapter 5, k-anonymity has been criticized for being ad hoc and not providing rigorous privacy semantics [33]. One of the major challenges in this field is in developing *practical* privacy frameworks which facilitate data access and interesting knowledge discovery in complex networks. Preserving privacy in dynamic social networks is also an important research direction which has received recent attention [14, 109].

PART III

Modeling, Evaluating, and Managing Users' Privacy Risk

In this part of the book, we shift the focus from the privacy issues that service providers face when providing access to their data, to the privacy-management concerns for the users of social-networking sites. More specifically, in this part, we give an overview of methods for modeling, evaluating and managing users' privacy risk. Chapter 7 describes a model which captures the tradeoff between the benefits of sharing and the risks of unwanted information propagation. Chapter 8 discusses methods for evaluating users' privacy risk, and Chapter 9 discusses the management of users' online privacy settings. Existing work on these aspects of privacy is currently limited and therefore, each chapter within Part III focuses on a particular aspect of user-specific privacy management and summarizes a representative work in the area.

Given the need for better control of one's online privacy, a growing number of collaborative sites allow individual users to set fine-grained policies that specify who can access their data and to what extent. We refer to such policies as the user's *privacy settings*. For example, Facebook provides a *Privacy Settings* page, which allows users to specify which pieces of their data (e.g., photos, personal information, etc.) will be visible by others. Since their introduction, Facebook has been trying to make these tools increasingly flexible and fine-grained [110]. Such tools allow individual users to construct and maintain *user-centric* policies governing access to their own content. In other words, users themselves define who will be able to access their data. For example, Facebook users define explicitly the visibility of their activity by their friends, friends-of-friends, or any Facebook user. This user-centric access control, in some sense, represents a paradigm shift from traditional approaches to access control where a professional system administrator defines and deploys access-control policies for an entire organization. Unfortunately, studies have consistently shown that it is difficult for most end-users to specify access-control policies for their own data [1, 26, 43, 63, 93]. As a result, it is common for users to ignore the privacy settings altogether [1, 43].

The exact relationship between users and their concern for their privacy still remains an open problem. For example, it has been revealed that, paradoxically, the more control users have over the publication of their private information, the more willing they are to publish sensitive information, even when the probability that strangers will access and use that information stays the same or, in fact, increases. On the other hand, less control over the publication of personal information increases individuals privacy concerns and decreases their willingness to publish sensitive information. This is true even when the probability that strangers will access and use that information actually decreases [20].

In this part of the book, we give an overview of privacy issues arising in social networks, viewed from the individual users' viewpoint. The related work in the area is young and not as mature as other existing privacy-related work. Therefore, the three chapters that compose this part of the book focus on extensively discussing new research in the area. We start by presenting recent game-theoretic studies [19, 56] that try to capture this trade-off between online social sharing and privacy risks. Then, we summarize proposed models and algorithms to measure the potential privacy risks for online users due to the information they share explicitly or implicitly. Finally, we discuss mechanisms that have been developed in order to help users better manage their privacy settings. In each part, we highlight some open problems and future directions of research.

CHAPTER 7

Models of Information Sharing

Users' participation in online social networks gives them a number of benefits, including staying in touch with friends and keeping informed about events and products. At the same time, their online presence also exposes them to the risks that their information may be propagated to people with whom they prefer not to share their information with.

In this chapter, we focus on presenting game-theoretic models that try to capture this tradeoff. More specifically, we focus on the recent game-theoretic model presented by Kleinberg and Ligett [56]. The main idea behind this work is to formalize the tradeoff between the benefit of sharing and the risk of exposure. In this game, every user is an individual who makes connections with other users in the network so as to maximize his/her utility; this utility encodes both the benefit and the risk factors. On a high level, the game studied by Kleinberg and Ligett is a *network-formation* game. In this game, users decide to make connections to others in a selfish manner. The main question that they answer is the following: "Given that users behave in a selfish way what will the network look like in an equilibrium state?" and "Will this network be stable?" In other words, the authors study whether there is a state in the network-formation game where users do not have any incentive to change their set of close friends with whom they share personal information. Their model builds upon the following sociology-theoretic intuitions.

1. Individuals benefit from information exchange; the ability of friends to learn things about each other strengthens their social tie.

2. All parties involved in the information exchange might leak information to strangers. That is, information spreads in the social network—to some extent—through gossip.

3. Information spread in the social network is both beneficial and harmful for the individual users. A user may receive positive utility from learning information about his friends and having friends learn information about him. However, there are also malevolent users. If personal information about an individual reaches such users, the former receives negative utility.

4. Therefore, a user decides to share information with a friend if the former believes that the information pathways will give him/her positive utility. A user avoids engaging in information-sharing activities that yield negative utility. A user may also decide to cut a link with one of his friends, if such a decision makes his other friends feel safer to share information with him/her. Every user computes the tradeoffs and makes decisions that yield a net personal utility benefit.

The formal definitions of a model that takes into account these high-level sociology principles will be described in the following sections.

7.1 INFORMATION-SHARING MODEL

The information-sharing model we describe here takes into account the sociological principles just described. Throughout this discussion, we assume a set of n users, denoted by V; some pairs of these users share personal information with each other (including indirect information that they have learned about others), and some pairs do not share personal information. Sharing of information is symmetric, and so if we let E denote the set of pairs who share information, then we obtain an information-sharing network $G = (V, E)$. Note that here we assume a simple setting where we ignore any information related to the affiliation of users, their attributes and their participation in groups.

If $i, j \in V$ are in the same connected component of G, then each user will learn personal information about the other, either by direct communication (if there is an (i, j) edge) or indirectly (if there is a path of length two or more from i to j). For any two users $i, j \in V$, user i receives a utility U_{ij} from being in the same component as j, and j receives a utility U_{ji} from being in the same component as i. These utilities can be either positive or negative, corresponding to the benefits of learning about your friends and being known to them, and the harms from having personal information reach users you are not friendly with. If $G(i)$ denotes the component of G containing i, then the total utility of i is equal to $\sum_{j \in G(i)} U_{ij}$. That is, the utility function of i depends on the size of the connected component that i belongs. Such a utility function is rather simplistic; in practice, a node does not directly benefit from all the nodes in the same component. Neither do all the nodes in the same component directly learn all the gossip about i. However, we will follow the methodology of Kleinberg and Ligett and focus on this simple case before introducing the more complicated scenarios they also discuss in their work.

7.2 STRATEGIC BEHAVIOR AND INFORMATION SHARING

The information-sharing model assumes that people mutually agree to share information. More specifically, each user is assumed to have one information item (i.e., one bit of information) that he/she wants to share. The model also implicitly assumes that information sharing maximizes users' utilities and therefore users are willing to share their bits with others. The utilities of the users are computed based on their expectations about what others will do. In other words, the assumption is that everybody in the set V knows the others and their connections.

Therefore, the setting represents a network-formation game, where each player must decide which links to maintain so as to maximize his/her utility, given the links everyone else has formed. The focus of the analysis is on studying the *stability* of the information-sharing network. In a stable network G, there is no incentive for players to add or drop links. In other words, users share information with the optimal set of other users, and in such a network users can trust that the information-sharing structure is self-enforcing. Note also that in contrast to many standard network-formation games, there is no explicit cost to maintain a link; the costs are implicit, based on the fact that a link exposes users to the risk that their information will reach other nodes with whom they might not be friends with.

Formally, the model proposed by Kleinberg and Ligett defines a defection from a formed network G to happen in either of the following two scenarios: (a) a single node i deletes a subset of its incident edges; or (b) a pair of nodes i, j agree to form the edge (i, j) and simultaneously delete subsets of their (other) incident edges. If there are no defections from G, in which all the participating nodes (i in case (a), and both i and j in case (b)) strictly improve their utilities, then the network is *stable*. Note that for defections of type (b), it is required that both i and j strictly improve their utilities.

Although pairwise interactions are fundamental to model information sharing, the model proposed by Kleinberg and Ligett can also be generalized to defections in which larger subsets of users coordinate their actions and cause a k-defection. That is, a k-defection consists of a set S of up to k nodes agreeing to form all pairwise edges within S, and simultaneously to each delete subsets of their (other) incident edges. Naturally, network G is *k-stable* if no k-defections are possible from G. In view of this general definition of defection, pairwise defections and pairwise stability are in fact a special case of the notion of k-defections and k-stability, where $k = 2$.

7.3 DISCUSSION AND SUMMARY OF RESULTS

Using the above model, Kleinberg and Ligett reach several interesting results, and identify interesting connections between their model and existing models of stability. We summarize these results below. *Symmetric utilities:* The simplest possible model is the one where utilities are symmetric (i.e., $U_{ij} = U_{ji}$ for every i, j) and take values from the set $\{-\infty, 1\}$. This model assumes that all pairs of users are either friends or enemies. In this case, there is a positive utility in sharing information with friends, but a much stronger negative utility in having malevolent users finding out personal information about you. The main result for this model, is that for any set of symmetric utilities from the set $\{-\infty, 1\}$, and every $k \geq 2$, a k-stable network always exists. For $k = 2$, which is the basic definition of stability, a stable network can be found in polynomial time. For general k, it is NP-hard to construct a k-stable network. For the intermediate case of fixed, constant $k > 2$, one can show how to construct k-stable networks in polynomial time for $k = 3$ and $k = 4$, while the investigation of the problem for larger constants k is left as an open question.

Since a socially optimal network may not be stable, one can also define the *price of stability*. This is the maximum welfare of any stable network relative to the optimum. Kleinberg and Ligett show that the price of stability is equal to 1 for 2-stable and 3-stable networks. This means that there always exist 1 and 2-stable networks that are also socially optimal. On the other hand, for $k > 3$ the price of stability exceeds 1. It is left as an open question to find a tight bound on the price of stability for $k > 3$.

The model with symmetric utilities becomes considerably harder, when the set of possible utility values expands from the set $\{-\infty, 1\}$ to the set $\{-\infty, 1, n\}$. This model can be used to capture scenarios where the friendly relations can be further partitioned into "weak ties" (that have weight 1) and "strong ties" (that have weight n) [42]. The relative values of the corresponding utilities are chosen so that the benefit of a single strong tie outweighs the total benefit of any number of weak ties incident to a single node. Kleinberg and Ligett point out that stable networks need not always exist in such a model. For this, they give the following example network, consisting of four people, who are related as follows: Anna has a strong tie to Bob; Claire has a strong tie to Daniel; Bob and Daniel are enemies; and other relations are weak ties. In any stable network, Anna and Bob would need to belong to the same component; Claire and Daniel would need to belong together in a different component. But then Anna and Claire would have an incentive to form an edge, violating stability. In fact, for any k, deciding whether a given instance contains a k-stable network is proved to be NP-complete.

Asymmetric utilities: The asymmetric-utilities case corresponds to a more general model in which for each pair of nodes i and j, node i receives a utility U_{ij} from being in the same component as j, and we may have $U_{ij} \neq U_{ji}$. On the positive side, this general model contains several other stability-related models as its special cases. For example, the Gale-Shapley Stable Marriage Problem with n men and n women [38] arises as a simple special case of the model. In order to see this, set $U_{ij} = -\infty$ for each pair of men and each pair of women. When i, j are of opposite genders, and i is in the n-th position in the i's preference list, then set $U_{ij} = 1 + n - p$. Related problems such as Becker's Marriage Game [11, 12] can be similarly reduced to the generalized model of Kleinberg and Ligett. At the same time, this generality gives rise to many more models, whose stability is unknown and can be direction for future research.

7.4 OPEN PROBLEMS

There are many possible generalizations of the above model. For example, it can be extended to handle different types of sensitive information, where each information type has different degree of sensitivity. Alternatively, information can be assumed to "attenuate" as it travels over multi-step paths. In practice, this means that information is being forgotten with some probability at each step, or that users decide with some probability not to propagate the information. Investigating how these factors can affect the above model are interesting directions of future research.

The model we described above, focuses on person-to-person communication. In OSNs, there are alternative means of information publishing. For example, in Facebook users publish information on their walls and anyone visiting the wall can see this information. Similarly, bloggers publish their views on their blogs and anyone other user has immediate access to them. The above model does not capture such publishing mechanisms. Neither does it capture cases where both person-to-person communication mechanisms are combined with alternative publishing techniques. Exploration of the notion of stability in such environments seems is an interesting research direction.

CHAPTER 8

Users' Privacy Risk

Although social-network users experience the benefits of their online presence, they are often unable to estimate the risks to their privacy imposed by their information-sharing activities. Even sophisticated users, who value privacy, appear to be willing to compromise their privacy in order to improve their digital presence in the virtual world. Users know that loss of control over their personal information poses a long-term threat, but they cannot assess the overall and long-term risk accurately enough to compare it to the short-term gain. Even worse, setting the privacy preferences in online services is often a complicated and time-consuming task that many users feel confused about and usually skip.

In this section, we summarize a mechanism that aims to provide social-network users with a feedback in the form of a score. This score, called *privacy score*, measures each user's potential privacy risk due to his/her online information-sharing behaviors. Using this score, a user can monitor his/her privacy risk in real time and compare it with the rest of the population to see where he/she stands. For example, a user can get informed that his/her privacy risk is higher than the privacy risk of his/her friends. This could alarm the user and guide him towards selecting more conservative privacy settings.

Overall, the privacy score of online social-network users can be thought of as an analogue of the credit score[1]. As the credit score quantifies the likelihood of an individual to default on a loan, the privacy score can be used as a measure of the tendency of a user to share his/her private information online. Therefore, high-level objective of a privacy score is to enhance public awareness of privacy, and to reduce the complexity of managing information sharing in social networks.

8.1 PRIVACY-SCORE MODEL

The *privacy score*, first introduced by Liu and Terzi [66], is a quantification of the privacy risk a user faces once he/she decides to participate in an online social network. The privacy score of a user is computed based on the user's information-sharing patterns. Intuitively, the more sensitive information a user shares and the more people he/she shares this information with, the higher the user's privacy risk; high privacy risk is also translated into high privacy score. In this paragraph, we give a formal definition of the privacy score as introduced by Liu and Terzi.

From the technical point of view, the definition of privacy score satisfies the following intuitive properties.

1. The score increases with the *sensitivity* of the information being revealed. That is, a user has high privacy score as more sensitive information about him/her is revealed.

2. The more *visible* the user's information becomes, the higher the privacy score of a user. That is, the larger number of people see the revealed information, the higher the privacy risk of the user.

The framework assumes that every user $j \in \{1, \ldots, n\}$ specifies his/her *privacy settings* for the same N profile items. These settings are stored in an $N \times n$ *response matrix* $\mathbf{R}$. The response matrix represents the profile privacy settings of user i. Value $\mathbf{R}(i, j)$ is an integer value that denotes how willing j is to disclose information about item i, with higher values to denote higher willingness. Also, higher values of $\mathbf{R}(i, j)$ entail at higher visibility of i. On the other hand, small values in the privacy settings of an item are an indication of high sensitivity; it is the highly-sensitive items that most of the people try to protect. Therefore, the response matrix $\mathbf{R}$ carries a lot of valuable information about users' privacy behavior.

For the rest of the discussion, we will adopt Liu and Terzi's notation and use β_i to denote the sensitivity of item i and Vis (i, j) to denote the visibility of the same item due to user j. The visibility value Vis (i, j) depends on the value $\mathbf{R}(i, j)$. However, this dependency is probabilistic. That is, Vis $(i, j) = \text{Prob} \{\mathbf{R}(i, j) = 1\}$, where Prob is computed assuming a particular probabilistic model. The role of such model is to encode the likelihood of entry (i, j) to be equal to 1, based on some assumptions about how users tend to set their privacy settings for different items. Different probabilistic models lead to different methods for computing the privacy score. In the next paragraphs, we present some of them and we discuss their advantages and disadvantages.

The privacy score of user j due to item i, denoted by $Pr(i, j)$ can be any combination of sensitivity and visibility. For example, by combining sensitivity and visibility using the product operator we get:

$$Pr (i, j) = \beta_i \times Vis (i, j).$$

Observe that $Pr (i, j)$ is monotonically increasing with both sensitivity and visibility.

In order to evaluate the overall privacy score of user j, denoted by $Pr (j)$, we can combine the privacy score of j due to different items. This can be done using, for example, the summation operation. In this case, the privacy score of user j is computed as follows:

$$Pr (j) \quad = \quad \sum_{i=1}^{N} Pr (i, j) = \sum_{i=1}^{N} \beta_i \times Vis (i, j). \qquad (8.1)$$

8.2 METHODS FOR COMPUTING THE PRIVACY SCORE

In this section, we discuss methods for computing the privacy score of user j, denoted by $Pr (j)$ and computed by Equation (8.1). We present two approaches: one is based on simple frequency computations, while the other is based on a more sophisticated method that users Item-Response Theory (IRT) [9]. For the rest of the discussion, we assume that the response matrix **R** is binary. That is, every user's preferences for an item is to either make it private or public. More complicated privacy settings are discussed towards the end of the section.

8.2.1 FREQUENCY-BASED METHOD

A simple model to compute the privacy score is the following naive model. If $|R_i|$ denotes the number of users who set $R (i, j) = 1$, then the sensitivity

β_i for binary matrices can be computed as the proportion of users that are reluctant to disclose item i. That is,

$$\beta_i = \frac{n - |\mathbf{R}_i|}{n}. \tag{8.2}$$

The higher the value of β_i, the more sensitive item i.

The computation of visibility in the binary case requires an estimate of the probability $P_{ij} = \text{Prob}\{\mathbf{R}(i, j) = 1\}$. Assuming independence between items and users, we can compute P_{ij} to be the product of the probability of an 1 in row $\mathbf{R}_i$ times the probability of an 1 in column $\mathbf{R}^j$. That is, if $|\mathbf{R}^j|$ is the number of items for which j sets $\mathbf{R}(i, j) = 1$, we have

$$P_{ij} = \frac{|\mathbf{R}_i|}{n} \times \frac{|\mathbf{R}^j|}{n}. \tag{8.3}$$

Probability P_{ij} is higher for less sensitive items and for users that have the tendency/attitude to disclose many of their profile items.

The naive computation of privacy score requires applying Equations (8.2) and (8.3) to Equation (8.1).

8.2.2 IRT-BASED METHOD

The IRT-based method computes the privacy score of a user using notions from the Item Response Theory (IRT) [9]. The advantages of IRT vs the naive definition of the privacy score will be discussed in detail later in this chapter. We only point out here that IRT encodes the sensitivity and visibility in an intuitive way. Moreover, experiments with real data demonstrate that the IRT model provides a better fit for the data. Last, but not least, the IRT model provides a privacy score that is medium-independent. That is, when the computations of privacy scores are done using the IRT-based method, the privacy scores of users across different networks are comparable. As we will discuss in more detail later, this is not true for the naive model. Before getting into the details of this adaptation, we give a brief introduction to IRT.

Introduction to IRT: IRT has its origins in educational statistics where it is used to analyze data from questionnaires and tests. The goal there is to measure the abilities of the examinees, the difficulty of the questions, and the probability of an examinee to correctly answer a given question.

The input consists of a set of n examinees and a set of N questions that they were asked. This input is put in an $N \times n$ response matrix $\mathbf{R}$, where $\mathbf{R}(i, j) = 1$ (resp. $\mathbf{R}(i, j) = 0$) if examinee j answered question i correctly (resp. wrongly). Given this input, the goal of IRT is to estimate the *difficulty* of each question and the *ability* of every examinee.

The IRT model that is most relevant to our discussion is the two-parameter model. In this model, every question q_i is characterized by a pair of parameters $\xi_i = (\alpha_i, \beta_i)$. Parameter β_i, $\beta_i \in (-\infty, \infty)$, represents the *difficulty* of q_i. Parameter α_i, $\alpha_i \in (-\infty, \infty)$, quantifies the *discrimination power* of q_i. Every examinee j is characterized by his/her ability level θ_j, $\theta_j \in (-\infty \infty)$. The basic random variable of the model is the response of examinee j to a particular question q_i. If this response is marked as either "correct" or "wrong" (dichotomous response), then the probability that j answers q_i **correctly** is given by

$$P_{ij} = \frac{1}{1 + e^{-\alpha_i(\theta_j - \beta_i)}}. \qquad (8.4)$$

Thus, P_{ij} is a function of parameters θ_j and $\xi_i = (\alpha_i, \beta_i)$. For a given question q_i with parameters $\xi_i = (\alpha_i, \beta_i)$, the plot of the above equation as a function of θ_j is called the *Item Characteristic Curve* (ICC).

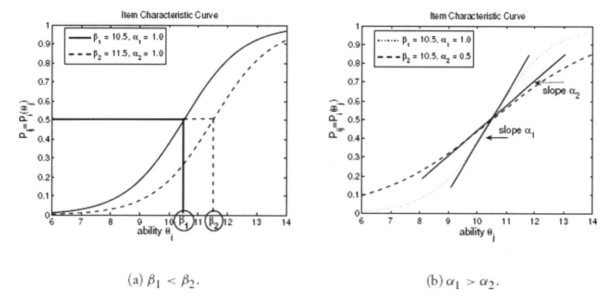

(a) $\beta_1 < \beta_2$.   (b) $\alpha_1 > \alpha_2$.

Figure 8.1: Item Characteristic Curves (ICC); y-axis: $P_{ij} = P_i(\theta_j)$ for different β values (Figure 8.1(a)) and α values (Figure 8.1(b)). x-axis: ability level θ_j.

The ICCs obtained for different values of parameters $\xi_i = (\alpha_i, \beta_i)$ are given in Figures 8.1(a) and 8.1(b). These figures make the intuitive meaning of parameters α_i and β_i easier to explain. Figure 8.1(a) shows the ICCs obtained for two questions q_1 and q_2 with parameters $\xi_1 = (\alpha_1, \beta_1)$ and $\xi_2 = (\alpha_2, \beta_2)$ such that $\alpha_1 = \alpha_2$ and $\beta_1 < \beta_2$. Parameter β_i, the item difficulty, is defined as the point on the ability scale at which $P_{ij} = 0.5$. We can observe that IRT places β_i and θ_j on the same scale (see the x-axis of Figure 8.1(a)) so that they can be compared. If an examinee's ability is higher than the difficulty of the question, then he/she has a better chance to get the answer right, and vice versa. This also indicates a very important feature of IRT called *group invariance*, that is, the item's difficulty is a property of the item itself, not of the people that responded to the item.

Figure 8.1(b) shows the ICCs obtained for two questions q_1 and q_2 with parameters $\xi_1 = (\alpha_1, \beta_1)$ and $\xi_2 = (\alpha_2, \beta_2)$ such that $\alpha_1 > \alpha_2$ and $\beta_1 = \beta_2$. Parameter α_i, the item discrimination, is proportional to the slope of $P_{ij} = P_i(\theta_j)$ at the point where $P_{ij} = 0.5$; the steeper the slope, the higher the discriminatory power of a question, meaning that this question can well differentiate among examinees whose abilities are below and above the difficulty of this question.

Mapping of IRT to privacy-score computation: In the IRT-based computation of the privacy score, the probability $\text{Prob}\{\mathbf{R}\ (i,\ j) = 1\}$ is estimated using Equation (8.4). However, in this case, there are no examinees and questions, but there are users and profile items. Thus, each examinee is mapped to a user, and each question is mapped to a profile item. The ability of an examinee corresponds to the *attitude* of a user: for user j, his/her attitude θ_j quantifies how concerned j is about his/her privacy; low values of θ_j indicate a conservative/introvert user, while high values of θ_j indicate a careless/extrovert user. The the difficulty parameter β_i is used to quantify the *sensitivity* of profile item i. In general, parameter β_i can take any value in $(-\infty, \infty)$. In order to maintain the monotonicity of the privacy score with respect to items' sensitivity one needs to guarantee that $\beta_i \geq 0$ for all $i \in \{1, \ldots, N\}$. This can be easily handled by shifting all items' sensitivity values by a large constant value.

In the above mapping, parameter α_i is ignored. Therefore, one could argue that it is enough to use the one-parameter model (q_i is only characterized by β_i), which is also known as the Rasch model.[2] In the Rasch model, each item is described by parameter β_i and $\alpha_i = 1$ for all $i \in \{1, \ldots, N\}$. However, as shown in [18] (and discussed in [9], Chapter 5), the Rasch model is unable to distinguish users that disclose the same number of profile items but with different sensitivities. On the other hand, the two-parameter model goes beyond counting and provides a finer-grained analysis of users attitude.

For computing the privacy score one first needs to compute the sensitivity β_i for all items $i \in \{1, \ldots, N\}$ and the probabilities $P_{ij} = \text{Prob}\{\mathbf{R}$ $(i,\ j) = 1\}$, using Equation (8.4). For the latter computation, we need to know all the parameters $\xi_i = (\alpha_i, \beta_i)$ for $1 \leq i \leq N$ and θ_j for $1 \leq j \leq n$. Liu and Terzi [66] estimate these parameters using as input the response matrix $\mathbf{R}$ and employing *Maximum Likelihood Estimation* (MLE) techniques.

More specifically, given parameters θ_j for $1 \leq j \leq N$, one can estimate the value of the parameters $\xi_i = (\alpha_i, \beta_i)$ for $1 \leq i \leq n$, using standard convex-optimization techniques (e.g., the Newton-Raphson [103] method). Similar techniques can be used to estimate the value of the parameters θ_j (for $1 \leq j \leq n$), if the values of $\xi_i = (\alpha_i, \beta_i)$ are known for all i's in $\{1, \dots, n\}$. Utilizing these techniques, one can use the *Expectation-Maximization* (EM) framework, in order to estimate the values of ξ_i and θ_j simultaneously, using as input solely the response matrix **R**. The sensitivity parameters β_i and the visibility P_{ij} computed using Equation (8.4) via the EM algorithm, can then be applied to Equation (8.1) in order to compute the privacy score of every user.

8.3 DISCUSSION AND SUMMARY OF RESULTS

The above framework can be further extended to non binary response matrices. More specifically, Liu and Terzi also consider matrices where **R** $(i, j) = k$, where $k \in \{1, \dots, \ell\}$. The interpretation of such values is the following: $\mathbf{R}(i, j) = k$ means that j discloses information related to item i to users that are at most k links away in a social network G. The authors adopt *polytomous* IRT models [9] in order to handle such cases in their framework.

As a means to validate the IRT framework for computing the privacy scores of users, Liu and Terzi conducted a survey on users' information-sharing preferences. Given a list of profile items that spanned a large spectrum of one's personal life (e.g., name, gender, birthday, political views, interests, address, phone number, degree, job, etc.), the users were asked to specify the extent they want to share each item with others. The privacy levels a user could allocate to items are $\{0, 1, 2, 3, 4\}$; **0** means that a user wants to share this item with *no one*; **1** means that he/she wants to share it with *some of his/her immediate friends*, **2** with *all of his/her immediate friends*, **3** with *all immediate friends and friends of friends*, and **4** with *everyone*. Along with the users privacy settings, Liu and Terzi have also collected information about the users location, educational background, age etc. The survey spanned 49 profile items and consisted of 153 complete responses from 18 countries/political regions. Among the participants, 53.3% were male and 46.7% were female, 75.4% were in the age of 23–39, 91.6% held a college degree or higher, and 76.0% reported to be spending 4 hours or more everyday surfing online. Some indicative experimental results from Liu and Terzi [66] are shown in Figure Figure 8.2. The tagcloud depicts the different profile items and their sensitivity values; the larger the fonts used to represent a profile item in the tag cloud, the higher its estimated sensitivity value. One can observe that *Mother's Maiden Name* is the most sensitive item, while one's *Gender*, which locates just right above the letter "h" of "Mother" has the lowest sensitivity; too small to be visually identified.

In a further extension of their work, Liu and Terzi also considered the structure of the users' social network in order to compute the visibility of items. More specifically, instead of using the value of the privacy setting per se to estimate the visibility of an item due to a user, they employee an information-propagation model so as to estimate the expected number of people that will know item i once it is revealed by user j at privacy level $R(i, j)$.

From a computational point of view, the naive computation can be done efficiently in $\mathcal{O}(nN)$ time. However, its disadvantage is that the sensitivity values obtained are significantly biased by the user population contained in R. As an example, consider two different social networks, e.g., LinkedIn and Facebook. As LinkedIn is a professional social network, people are more conservative and they are reluctant to reveal details about their personal lives to their professional friends. On the other hand, friendships in Facebook are more informal and Facebook users are usually more open and willing to reveal some more personal information (e.g., their party photos) to their Facebook friends. Using the naive privacy-risk model, the sensitivity values of items in the conservative population of LinkedIn users will be very high, causing the naive privacy scores of LinkedIn users to have higher values than the naive privacy scores of Facebook users. This is not only counter-intuitive, but it also reveals the disadvantage of the naive model; namely the scores are not *medium invariant*. In other words, the naive scores obtained in two different networks are not comparable with each other. This is not true for the IRT-based privacy score. The IRT-based scores of users are comparable across social networks. Therefore, independently of the nature of the network and the sharing behavior of users within it, the scores computed for the users of one network are comparable to the scores computed for the users in another network.

Figure 8.2: Sensitivity of the profile items computed using IRT model with input the dichotomous matrix R_2^*. Larger fonts mean higher sensitivity.

Overall, the advantages of the IRT framework can be summarized as follows.

(1) The quantities IRT computes, i.e., sensitivity, attitude and visibility, have an intuitive interpretation. For example, the sensitivity of information can be used to send early alerts to users when the sensitivities of their shared profile items are out of the comfortable region.

(2) Due to the independence assumptions, many of the computations can be parallelized, which makes the computation very efficient in practice.

(3) The experimental results by Liu and Terzi demonstrate that the probabilistic model defined by IRT in Equation (8.4) can be viewed as a generative model, and it fits the real response data very well in terms of χ^2 goodness-of-fit test.

(4) Most importantly, the estimates obtained from IRT framework satisfy the *group invariance* property. At an intuitive level, this property means that the sensitivity of the same profile item estimated from different social networks are close to the same true value, and consequently, the privacy scores computed across different social networks are comparable.

8.4 OPEN PROBLEMS

All the techniques described above exploit the following three independence assumptions that are inherent in IRT models: (i) independence between items; (ii) independence between users; and (iii) independence between users and items. These independence assumptions are necessary for devising a relatively simple and intuitive model. Also, they help in the design of efficient algorithms for computing the privacy scores. However, in practice, many items are correlated as is the behavioral patterns of many users. Incorporating such dependencies is an interesting direction towards an even more realistic model.

Even further, the privacy-risk model we described in this section provides the users with feedback in the form of a single privacy score. While a single score is easy to process and comprehend, capturing a user's privacy settings using a single number may hide a lot of useful details from the user. Thus, instead of a single score, one can also explore the option of showing t multiple (but not too many) scores to each user. One of the main challenges is to find a balance between detailed and concise feedback in a formal and automatic manner.

Furthermore, a user participating in an online service might be associated with a set of a-priori defined groups. Each such group g contains a subset of the users and it thus defines a response matrix $\mathbf{R}_g$; $\mathbf{R}_g$ is a submatrix of $\mathbf{R}$ containing only the settings of the users in g. For computing the score of a user within group g, one can use any of the above methodologies. For example, a user participating in groups `Purdue alumni` and `BU Faculty` will have a separate privacy score computed based on the responses of the users in each group.

There may be cases, however, where the groups are not a-priori defined. Hence, a future research direction is the design of methods that can automatically find interesting submatrices of the response matrix **R** and compute the privacy scores of the users based on the information in these submatrices. This partitioning of the submatrices seems related to the notion of the *co-clustering* [7] (also known as *bi-clustering* [83]). Of course, the objective function here is different: the rows and the columns of the global response matrix **R** need to be partitioned, so that each submatrix is best described by a model (e.g., an IRT-based model described by Equation (8.4)). Further, the number of the rows and columns of **R** will be split on, need to be determined automatically. One approach is to use the Minimum Description Length (MDL) principle [86] in order to determine the right number of privacy scores that will be associated with every user. Note that in this setting, all users will not be associated with the same number of scores. Rather, the number of scores associated with user j can depend on the settings of user j as well as the settings of other users.

[1] www.myfico.com

[2] http://en.wikipedia.org/wiki/Rasch_model

CHAPTER 9

Management of Privacy Settings

Despite the well-documented privacy risks for users of online social networks, it has been consistently shown that end users struggle to express and maintain detailed policies controlling access to their data [1, 26, 43, 63, 93]. This is due, in part, to user interfaces that are complex and hard to use [93]. On Facebook, for example, the user must manually assign friends to lists. As the average Facebook user has 130 friends[1], the process can be very time-consuming. Worse, numerous lists may be required since a user's privacy preferences can be different for different personal-information items (e.g., Home Address vs. Religious Views).

In the previous chapter, we described mechanisms that raise users' awareness with respect to their tendency to share information in their virtual lives. These information-raising mechanisms were in the form of alerts given to the users. In this chapter, we describe existing work that aims towards the development of tools that will help users adjust their privacy settings. Such tools take as input a subset of each user's privacy settings and make user-specific recommendations for the rest of the user's privacy settings. That is, these tools go beyond alerts and actually help the users to be proactive and guide them towards changing their settings. For example, based on the fact that 20-year old undergraduates share their photos with their other 20-year-old friends, but not with their professors, such a tool can make recommendations about the photos privacy settings to a 20-year-old user. We call such tools *privacy wizards*. One should think of privacy wizards as mechanisms that use data-mining and machine-learning techniques as well as limited user input, in order to learn a *privacy-preference model* for each user. Using this model, the privacy wizard can automatically recommend privacy settings to every user separately.

We devote the next paragraphs describing the privacy wizard that was developed by Fang and LeFevre [37]. In order to learn an appropriate preference model, the wizard solicits some input from the user. This is done by asking the users specific questions. Hence, the prototype wizard solicits input from the user by asking her to assign labels (i.e., *yes* or *no*) to particular *(data item, friend)* pairs. For example, a standard interaction asks "Would you like to share *Date of Birth* with *Alice Smith*?" It is widely accepted that users (particularly non-technical users) have difficulty reasoning holistically about privacy and security policies [63, 85]. Thus, such example-based questions are much more effective in capturing users' information-sharing intentions.

Of course, the user's attention is limited. A naive approach might ask the user to exhaustively label all of his/her friends; but since the average Facebook user has more than 100 friends, this is impractical. Instead, it is critical that the wizard *asks the right questions*, or selects the *most informative* friends for the user to label. Given such a clever data acquisition strategy, the goal of the wizard, then, is to use the labeled training data (friends) to construct a model, and to use this model to automatically recommend the rest of the user's privacy settings. We discuss the details of this paradigm next.

9.1 A MODEL FOR MANAGING PRIVACY SETTINGS

Following the model proposed by Fang and LeFevre [37], we consider again the response matrix $\mathbf{R}$. In this case, there are again n users and N attributes. Using the notation of the previous chapter, the j-th row of the response matrix $\mathbf{R}$ corresponds to the privacy settings of user j. More specifically, the notation used in the previous chapter indicates that $\mathbf{R}(i, j) = 1$ (resp. $\mathbf{R}(i, j) = 0$) implies that user j indicated that his/her information with respect to profile item i is set to be public (resp. private). In the work of Fang and LeFevre [37], the authors add one more additional level of detail in the response matrix $\mathbf{R}$. That is, the privacy settings of the j-th user with respect to item i are encoded in an n dimensional vector $\mathbf{R}(i, j)[]$, where $\mathbf{R}(i, j)[u] = 1$ (resp. $\mathbf{R}(i, j)[u] = 0$) denotes that user j has set attribute i visible (resp. invisible) to user u. Therefore, entries $\mathbf{R}(i, j)[u]$ for every item i and every user u represent the privacy settings of user j.

The goal of the privacy wizard proposed by Fang and LeFevre [37] is to build a model and a system that can help user j automatically set the entries $\mathbf{R}(i, j)[u]$ for every item i and user u, by asking j to manually set the smallest possible number of privacy settings. That is, every user is asked to specify his/her choice of privacy settings with respect to a small subset of their friends. The wizard is then used to predict the rest of the entries $\mathbf{R}(i, j)$ $[u]$.

For this prediction task, the authors use machine-learning techniques. That is, the privacy wizard is a classifier that predicts the value of $\mathbf{R}(i, j)[u]$ as accurately as possible, using the smallest possible number of training examples. In this case, the training examples correspond to the number of privacy settings that user j will be asked to set manually. In order to achieve this, Fang and LeFevre adopt an *active learning* approach.

The classifier built by Fang and LeFevre uses two types of features to characterize each user.

- *Community structure features*. For these features, all the friends of a user that belong in a specific community (identified by a community-detection algorithm) are grouped together. The method assumes that if a particular attribute is made visible to one member of the community, then—with high probability—it is also made visible to the other members of the community too. In practice, communities are assumed to be densely connected subgraphs defined in the social-network graph G.

- *Other profile information*. Other features that are taken into account in order to build the classifier have to do with demographic information associated with the users' profiles; Gender, Age, Education, Relationship Status, Political, and Religious Views being some of them. In addition to these demographic information, this type of features might also include the participation of the user in groups, "fan" pages events or even tagged photos.

The extracted features are used to build the *feature matrix* associated with every user j.

For example, assume user j that has 3 friends and has been identified to belong in two communities in the graph, C_1 and C_2. Additionally assume that user j is connected with three other people, *Alice, Bob* and *Carol*. Now assume that for a particular profile item, i.e., date of birth (DOB), j has decided to reveal it to Alice and Bob but not to Carol. This would mean that $\mathbf{R}(DOB, j)[Alice] = 1$, $\mathbf{R}(DOB, j)[Bob] = 1$ and $\mathbf{R}(DOB, j)[Carol] = 0$. Additionally, assume that Alice and Bob participate in both communities C_1 and C_2 that j also participates in, while Carol participates only in community C_1. Table 9.1 shows an example of the combined feature and response matrix of user j for item DOB.

Table 9.1: An example of a combined feature and response matrix for user j for attribute DOB.

	Age	Gender	Obama fan	C_1	C_2	$R(DOB, j)[u]$
Alice	25	F	Yes	1	1	1
Bob	38	M	No	1	1	1
Carol	22	F	Yes	1	0	0

9.2 PREDICTING USERS' PRIVACY SETTINGS

The privacy wizard works in two steps. For every user j, it first identifies a set of j's friends that is going to explicitly use in order to ask j whether he/she wants to share certain attributes with these friends. In the second phase, the wizard forms the profile of user j (i.e., it builds a classifier), which is then used to predict the rest of the entries of $\mathbf{R}(i, j)[u]$ for every u who is a friend of j and every profile item i.

Of course, the performance of privacy wizard depends on two factors (a) the number of friends that the user is asked to declare his/her privacy settings, and (b) the accuracy of the classifier built using the declared privacy settings and the extracted features. Since the amount of effort that the user is willing to devote in declaring his/her privacy settings is limited and unpredictable, the goal is to build a classifier that is able to have high accuracy even with small amount of training data.

In order to achieve the best accuracy possible, with the understanding that the user may quit giving information about what he/she wants to share with his/her friends at any time, Fang and LeFevre adopted the active learning paradigm known as *uncertainty sampling* [59]. The uncertainty sampling paradigm guides the privacy wizard to operate in rounds. At every round, the wizard selects a subset of a users' friends and ask him/her to declare his/her privacy settings with respect to these friends. Based on the user's answers the classifiers builds a model using these responses and the feature matrix associated with this subset of the user's friends. The key characteristic of the algorithm is in the way it picks the set of friends that the user needs to label by selecting those that minimize the entropy of the predictions on the yet-unlabeled set of friends.

In principle, the uncertainty sampling paradigm allows the privacy wizard to decide when it has enough information, so that more labels from this user will not improve the predictive power of the wizard. Nevertheless, in practice, it can be the case that the user stops giving his/her privacy settings before the wizard has enough information to make accurate predictions.

9.3 DISCUSSION AND SUMMARY OF RESULTS

Fang and LeFevre [37] have implemented their approach within a Facebook application and used it to conduct a Facebook-user survey and evaluate the practical utility of the wizard. More specifically, their experimental evaluation focused both on checking the accuracy of the wizard in predicting the users' settings and on quantifying the importance of different features (community structure or demographic information) in the classification accuracy. The results indicate that the privacy wizard that uses the uncertainty sampling technique has high accuracy in predicting the privacy settings of Facebook users. As far as feature importance is concerned, their experiments reveal that community-based features are more important in determining the choice of a user to share or not some of his/her personal information.

The survey results of Liu and Terzi [66] showed that people from North America and Europe are much more willing to share their private information online. One possible explanation for this is that people from North America and Europe are more comfortable in revealing personal information on the social networks they participate. Since online social-networking is more widespread in these regions, one can assume that people in North America and Europe succumb to the social pressure to reveal things about themselves online in order to appear "cool" and become popular among their friends that also participate in the network. The role of the community in determining users' privacy settings has also been revealed in the study of Fang and LeFevre [37]. In this latter case, the connectivity-based features have appeared to have higher importance, i.e., they were able to predict the tendency of users to share or not to share their private information.

These observations are indicative that the users of online social networks may not be aware of how much they actually reveal online. People are more willing to reveal information about themselves online, than in their real lives. Therefore, mechanisms like the privacy score will increase awareness of users towards what they are actually doing by selecting a specific configuration of their privacy settings. Privacy wizards will further help users define their privacy settings so that they do not leave some aspects of their private lives unprotected. A combination mechanism that can incorporate privacy-score information within privacy wizards is an interesting direction of future research.

9.4 OPEN PROBLEMS

The privacy wizard is an important step towards user studies that allow us to better understand and model what users are willing to share in a social setting and what are the factors that affect their decisions. Facebook, by enabling Facebook applications, provides a nice platform where such studies can be conducted. Such studies can prove extremely useful in understanding human behavior and the attitude of different social groups towards their privacy.

At the same time, the emergence of Facebook, Google+, Flickr and many other types of social media propose a paradigm shift from systems where the access control mechanisms were designed and implemented by experts (e.g., system administrators) to an era where unspecialized users need to decide, design and implement their privacy settings for their personal information. Clearly, the privacy wizard is a useful tool that can assist non-specialized users towards achieving the hard goal of protecting their personal data. The privacy wizard proposed by Fang and LeFevre provide a nice framework where suggestions for privacy settings can be automatically provided within Facebook. How similar mechanisms can be materialized in different OSNs (e.g., Google+, Flickr, etc.) is a direction for future work.

Privacy wizards, like the one proposed by Fang and LeFevre [37] are a new type of interfaces that allow for the interaction of users with social media. The functionality and the usability of wizards open up interesting questions in the area of human-computer interaction and interface design. For example, in a more recent work Mazzia et al. [73] propose how to build interfaces that will allow untrained users to *understand* their privacy settings and their implications. Different types of social media, need different types of interfaces. This can be not only because the media are used by different groups of people; for example some types of networks are only joined by professionals and technocrats who have the technological know-how, while others are joined by a wider set of people. Therefore, privacy-management interfaces should take the user community into account. Similarly, the type of the social media itself, might require different types of interfaces. Overall, the privacy wizard we described above opens up a new direction of research in the design of interfaces that such media need to have in order to be effective and efficient for the different communities of candidate users.

---

[1] http://www.facebook.com/press/info.php?statistics

Bibliography

[1] A. Acquisti and R. Gross. Imagined communities: Awareness, information sharing, and privacy on the facebook. In *Privacy Enhancing Technologies Workshop*, 2006. DOI: 10.1007/11957454_3 Cited on page(s) 44, 59

[2] L. Adamic and E. Adar. Friends and neighbors on the web. *Social Networks*, 25(3):211–230, 2003. DOI: 10.1016/S0378-8733(03)00009-1 Cited on page(s) 20

[3] C. Aggarwal, editor. *Social Network Data Analytics*. Springer, 2011. DOI: 10.1007/978-1-4419-8462-3 Cited on page(s) 1

[4] C. C. Aggarwal and P. S. Yu. *Privacy-Preserving Data Mining: Models and Algorithms*. Springer, 2008. DOI: 10.1007/978-0-387-70992-5 Cited on page(s) 3

[5] G. Aggarwal, T. Feder, K. Kenthapadi, R. Motwani, R. Panigrahy, D. Thomas, and A. Zhu. Approximation algorithms for k-anonimity. *Journal of Privacy Technology*, November 2005. DOI: 10.1145/1247480.1247490 Cited on page(s) 26

[6] E. Airoldi, D. Blei, S. Fienberg, and E. Xing. Mixed-membership stochastic blockmodels. *Journal of Machine Learning Research (JMLR)*, 9:1981–2014, 2008. Cited on page(s) 20

[7] A. Anagnostopoulos, A. Dasgupta, and R. Kumar. Approximation algorithms for co-clustering. In *ACM Symposium on Principles of Database Systems (PODS)*, pages 201–210, 2008. DOI: 10.1145/1376916.1376945 Cited on page(s) 58

[8] L. Backstrom, C. Dwork, and J. Kleinberg. Wherefore art thou r3579x: anonymized social networks, hidden patterns, and struct

steganography. In *International World Wide Web Conference (WWW)*, 2007. DOI: 10.1145/1242572.1242598 Cited on page(s) 12, 15, 33

[9] F. B. Baker and S.-H. Kim. *Item Response Theory: Parameter Estimation Techniques*. Marcel Dekkerm, Inc., 2004. Cited on page(s) 52, 53, 55, 56

[10] M. Barbaro and T. Zeller. A face is exposed for aol searcher no. 4417749. *New York Times*, August 2006. Cited on page(s) 16

[11] G. Becker. A theory of marriage: Part I. *Journal of Political Economy*, 81(4):813, 1973. DOI: 10.1086/260084 Cited on page(s) 50

[12] G. Becker. A theory of marriage: Part II. *Journal of Political Economy*, 81(S2):11, 1974. Cited on page(s) 50

[13] S. Bhagat, G. Cormode, B. Krishnamurthy, and D. Srivastava. Class-based graph anonymization for social network data. In *International Conference on Very Large Databases (VLDB)*, 2009. Cited on page(s) 15, 36, 41

[14] S. Bhagat, G. Cormode, B. Krishnamurthy, and D. Srivastava. Privacy in dynamic social networks. In *International World Wide Web Conference (WWW)*, 2010. DOI: 10.1145/1772690.1772803 Cited on page(s) 42

[15] S. Bhagat, G. Cormode, and S. Muthukrishnan. Node classification in social networks. In C. Aggarwal, editor, *Social Network Data Analytics*, pages 115–148. Springer, 1 edition, 2011. DOI: 10.1007/978-1-4419-8462-3_5 Cited on page(s) 18, 19

[16] I. Bhattacharya and L. Getoor. Collective entity resolution in relational data. *ACM Transactions on Knowledge Discovery from Data*, 1(1):1–36, March 2007. DOI: 10.1145/1217299.1217304 Cited on page(s) 17

[17] L. Bilge, T. Strufe, D. Balzarotti, and E. Kirda. All your contacts are belong to us: Automated identity theft attacks on social networks. In *International World Wide Web Conference (WWW)*, 2009. DOI: 10.1145/1526709.1526784 Cited on page(s) 12

[18] A. Birnbaum. Some latent trait models and their use in inferring the examinee's ability. *Statistical theories of mental test scores*, pages 397–479, 1968. Cited on page(s) 55

[19] L. Blume, D. Easley, J. Kleinberg, R. Kleinberg, and E. Tardos. Network formation in the presence of contagious risk. In *ACM Conference on Electronic Commerce*, 2011. DOI: 10.1145/1993574.1993576 Cited on page(s) 44

[20] L. Brandimarte, A. Acquisti, and G. Loewenstein. Misplaced confidences: Privacy and the control paradox, 2010. Cited on page(s) 44

[21] G. Brown, T. Howe, M. Ihbe, A. Prakash, and K. Borders. Social networks and context-aware spam. In *ACM Conference on Computer Supported Collaborative Work*, 2008. DOI: 10.1145/1460563.1460628 Cited on page(s) 1

[22] A. Campan and T. M. Truta. A clustering approach for data and structural anonymity in social networks. *KDD Workshop on Privacy, Security, and Trust in KDD (PinKDD)*, 2008. Cited on page(s) 12, 13, 33, 36, 37

[23] S. Chawla, C. Dwork, F. Mcsherry, A. Smith, and H. Wee. Toward privacy in public databases. In *Theory of Cryptography Conference (TCC)*, 2005. DOI: 10.1007/978-3-540-30576-7_20 Cited on page(s) 25

[24] B.-C. Chen, D. Kifer, K. LeFevre, and A. Machanavajjhala. Privacy-preserving data publishing. *Foundations and trends in databases*, 2(1–2):1–167, 2009. DOI: 10.1561/1900000008 Cited on page(s) 3

[25] D. R. Choffnes, J. Duch, D. Malmgren, R. Guimera, F. E. Bustamante, and L. Amaral. Swarmscreen: Privacy through plausible deniability in p2p systems tech. Technical Report NWU-EECS-09-04, Department of EECS, Northwestern University, June 2009. Cited on page(s) 16

[26] L. Church, J. Anderson, J. Bonneau, and F. Stajano. Privacy stories: Confidence on privacy behaviors through end user programming. In *Symposium on Usable Privacy and Security (SOUPS)*, 2009. DOI: 10.1145/1572532.1572559 Cited on page(s) 44, 59

[27] K. Clarkson, K. Liu, and E. Terzi. Towards identity anonymization in social networks. In P. Yu, J. Han, and C. Faloutsos, editors, *Link Mining: Models Algorithms and Applications*. Springer, 2010. DOI: 10.1007/978-1-4419-6515-8 Cited on page(s) 3

[28] G. Cormode, D. Srivastava, T. Yu, and Q. Zhang. Anonymizing bipartite graph data using safe groupings. In *International Conference on Very Large Databases (VLDB)*, 2008. DOI: 10.1145/1453856.1453947 Cited on page(s) 16, 28, 40, 41

[29] C. Cortes, D. Pregibon, and C. Volinsky. Communities of interest. In *Advances in Intelligent Data Analysis*, 2001. DOI: 10.1007/3-540-44816-0_11 Cited on page(s) 16

[30] danah boyd. Privacy and publicity in the context of big data. In *International World Wide Web Conference (WWW)*, 2010. Invited talk. Available at http://www.danah.org/papers/talks/2010/WWW2010.html. Cited on page(s) 3

[31] S. Das, Ö. Eğecioğlu, and A. E. Abbadi. Anonymizing weighted social network graphs. In *IEEE International Conference on Data Engineering (ICDE)*, 2010. DOI: 10.1109/ICDE.2010.5447915 Cited on page(s) 15

[32] C. Dwork. Differential privacy. In *International Colloquium on Automata, Languages and Programming (ICALP)*, 2006. Cited on page(s) 3, 29

[33] C. Dwork. An ad omnia approach to defining and achieving private data analysis. *KDD Workshop on Privacy, Security, and Trust in KDD (PinKDD) 2007*, 4890:1–13, 2007. DOI: 10.1007/978-3-540-78478-4_1 Cited on page(s) 29, 42

[34] C. Dwork, K. Kenthapadi, F. McSherry, I. Mironov, and M. Naor. Our data, ourselves: privacy via distributed noise generation. In *International Conference on the Theory and Applications of Cryptographic Techniques (EUROCRYPT)*, 2006. DOI: 10.1007/11761679_29 Cited on page(s) 30

[35] C. Dwork, F. McSherry, K. Nissim, and A. Smith. Calibrating noise to sensitivity in private data analysis. In *Theory of Cryptography Conference (TCC)*, 2005. DOI: 10.1007/11681878_14 Cited on page(s) 30

[36] D. Easley and J. Kleinberg, editors. *Networks, Crowds, and Markets*. Cambridge University Press, 2010. Cited on page(s) 1

[37] L. Fang and K. LeFevre. Privacy wizards for social networking sites. In *International World Wide Web Conference (WWW)*, 2010. DOI: 10.1145/1772690.1772727 Cited on page(s) 59, 60, 62, 63

[38] D. Gale and L. Shapley. College admissions and the stability marriage. *The Americal Mathematical Monthly*, 69(1):9–15, 1962. DOI: 10.2307/2312726 Cited on page(s) 50

[39] L. Getoor and C. P. Diehl. Link mining: a survey. *SIGKDD Explorations Newsletter*, 7(2):3–12, December 2005. DOI: 10.1145/1117454.1117456 Cited on page(s) 17

[40] A. Goldenberg, A. Zheng, S. Fienberg, and A. E.M. A survey of statistical network models. In *Foundations and Trends in Machine*

Learning, volume 2, pages 129–233. Now Publishers, 2009. DOI: 10.1561/2200000005 Cited on page(s) 20

[41] J. Gomez, T. Pinnick, and A. Soltani. Knowprivacy, June 2009. Cited on page(s) 3

[42] M. Granovetter. The strength of weak ties. *American Journal of Sociology*, 78:1360–1380, 1973. DOI: 10.1086/225469 Cited on page(s) 50

[43] R. Gross and A. Acquisti. Information revelation and privacy in online social networks. In *Workshop on Privacy in the Electronic Society*, 2005. DOI: 10.1145/1102199.1102214 Cited on page(s) 44, 59

[44] M. Hasan, V. Chaoji, S. Salem, and M. Zaki. Link prediction using supervised learning. In *SDM Workshop on Link Analysis, Counterterrorism and Security*, 2006. Cited on page(s) 20

[45] M. A. Hasan and M. J. Zaki. A survey of link prediction in social networks. In C. Aggarwal, editor, *Social Network Data Analytics*, pages 243–276. Springer, 1 edition, 2011. DOI: 10.1007/978-1-4419-8462-3 Cited on page(s) 19

[46] M. Hay, M. Hay, V. Rastogi, G. Miklau, and D. Suciu. Boosting the accuracy of differentially-private histograms through consistency. In *International Conference on Very Large Databases (VLDB)*, 2010. Cited on page(s) 31

[47] M. Hay, C. Li, G. Miklau, and D. Jensen. Accurate estimation of the degree distribution of private networks. In *IEEE International Conference on Data Mining (ICDM)*, 2009. DOI: 10.1109/ICDM.2009.11 Cited on page(s) 31, 36

[48] M. Hay, K. Liu, G. Miklau, J. Pei, and E. Terzi. Privacy-aware data management in information networks. In *SIGMOD Conference*,

pages 1201–1204, 2011. DOI: 10.1145/1989323.1989453 Cited on page(s) 3

[49] M. Hay, G. Miklau, and D. Jensen. Enabling accurate analysis of private network data. In F. Bonchi and E. Ferrari, editors, *Privacy-Aware Knowledge Discovery: Novel Applications and New Techniques*. Chapman & Hall/CRC Press, 2010. DOI: 10.1201/b10373 Cited on page(s) 3

[50] M. Hay, G. Miklau, D. Jensen, and D. Towsley. Resisting structural identification in anonymized social networks. In *International Conference on Very Large Databases (VLDB)*, August 2008. DOI: 10.1145/1453856.1453873 Cited on page(s) 12, 33, 34, 35, 36

[51] M. Hay, G. Miklau, D. Jensen, P. Weis, and S. Srivastava. Anonymizing social networks. Technical report, University of Massachusetts, Amherst, March 2007. Cited on page(s) 12, 28, 33, 34, 35, 36

[52] M. Hernandez and S. Stolfo. The merge/purge problem for large databases. In *SIGMOD*, 1995. DOI: 10.1145/568271.223807 Cited on page(s) 17, 18

[53] T. N. Herzog, F. Scheuren, and W. Winkler, editors. *Data quality and record linkage techniques*. Springer, 2007. Cited on page(s) 17

[54] D. Kifer and A. Machanavajjhala. Personalized social recommendations - accurate or private? In *ACM SIGMOD International Conference on Management of Data (SIGMOD)*, 2011. Cited on page(s) 30, 31

[55] J. M. Kleinberg. Challenges in mining social network data: processes, privacy, and paradoxes. In *ACM SIGKDD International Conference on Knowledge Discovery and Data Mining (KDD)*, pages 4–5, 2007. DOI: 10.1145/1281192.1281195 Cited on page(s) 3

[56] J. M. Kleinberg and K. Ligett. Information-sharing and privacy in social networks. *CoRR*, abs/1003.0469, 2010. DOI: 10.1145/1281192.1281195 Cited on page(s) 44, 47

[57] A. Korolova, K. Kenthapadi, N. Mishra, and A. Ntoulas. Releasing search queries and clicks privately. In *International World Wide Web Conference (WWW)*, 2009. DOI: 10.1145/1526709.1526733 Cited on page(s) 12, 40

[58] A. Korolova, R. Motwani, S. U. Nabar, and Y. Xu. Link privacy in social networks. In *ACM Conference on Information and Knowledge Management (CIKM)*, 2008. DOI: 10.1145/1458082.1458123 Cited on page(s) 15

[59] D. D. Lewis and W. A. Gale. A sequential algorithm for training text classifiers. In *SIGIR*, pages 3–12, 1994. DOI: 10.1145/219587.219592 Cited on page(s) 61

[60] N. Li, T. Li, and S. Venkatasubramanian. t-closeness: Privacy beyond k-anonymity and l-diversity. In *IEEE International Conference on Data Engineering (ICDE)*, 2007. DOI: 10.1109/ICDE.2007.367856 Cited on page(s) 28, 29

[61] D. Liben-Nowell and J. Kleinberg. The link prediction problem for social networks. In *ACM Conference on Information and Knowledge Management (CIKM)*, 2003. DOI: 10.1002/asi.v58:7 Cited on page(s) 20

[62] J. Lindamood, R. Heatherly, M. Kantarcioglu, and B. Thuraisingham. Inferring private information using social network data. In *International World Wide Web Conference (WWW)*, 2009. DOI: 10.1145/1526709.1526899 Cited on page(s) 13, 41

[63] H. Lipford, A. Besmer, and J. Watson. Understanding privacy settings in facebook with an audience view. In *Conference on Usability, Psychology, and Security*, 2008. Cited on page(s) 44, 59

[64] K. Liu, K. Das, T. Grandison, and H. Kargupta. Privacy-preserving data analysis on graphs and social networks. In H. Kargupta, J. Han, P. Yu, R. Motwani, and V. Kumar, editors, *Next Generation of Data Mining*, chapter 21, pages 419–437. Chapman & Hall/CRC, 2008. DOI: 10.1201/9781420085877 Cited on page(s) 3

[65] K. Liu and E. Terzi. Towards identity anonymization on graphs. In *ACM SIGMOD International Conference on Management of Data (SIGMOD)*, 2008. DOI: 10.1145/1376616.1376629 Cited on page(s) 12, 28, 33, 34, 35, 37

[66] K. Liu and E. Terzi. A framework for computing the privacy scores of users in online social networks. *TKDD*, 5(1):6, 2010. DOI: 10.1145/1870096.1870102 Cited on page(s) ix, 51, 55, 56, 62

[67] L. Liu, J. Wang, J. Liu, and J. Zhang. Privacy preservation in social networks with sensitive edge weights. In *SIAM International Conference on Data Mining (SDM)*, 2009. Cited on page(s) 15

[68] L. Lü and T. Zhou. Link prediction in complex networks: A survey. *Physica A: Statistical Mechanics and its Applications*, 390(6):1150 – 1170, 2011. DOI: 10.1016/j.physa.2010.11.027 Cited on page(s) 19

[69] Q. Lu and L. Getoor. Link based classification. In *International Conference on Machine Learning (ICML)*, 2003. Cited on page(s) 18

[70] A. Machanavajjhala, J. Gehrke, D. Kifer, and M. Venkitasubramaniam. l-diversity: Privacy beyond k-anonymity. In *IEEE International Conference on Data Engineering (ICDE)*, 2006. DOI: 10.1145/1217299.1217302 Cited on page(s) 27, 28

[71] A. Machanavajjhala, A. Korolova, and A. D. Sarma. Personalized social recommendations - accurate or private? In *International Conference on Very Large Databases (VLDB)*, 2011. Cited on page(s) 19, 37

[72] S. Macskassy and F. Provost. Classification in networked data: A toolkit and a univariate case study. *Journal of Machine Learning Research (JMLR)*, 8:935–983, May 2007. Cited on page(s) 19

[73] A. Mazzia, K. LeFevre, and E. Adar. A tool for privacy comprehension. In *CHI Workshops*, 2011. Cited on page(s) 63

[74] A. McCallum, K. Nigam, and L. Ungar. Efficient clustering of high-dimensional data sets with application to reference matching. In *KDD*, 2000. DOI: 10.1145/347090.347123 Cited on page(s) 18

[75] L. McDowell, K. Gupta, and D. Aha. Cautious inference in collective classification. In *AAAI Conference on Artificial Intelligence (AAAI)*, 2007. DOI: 10.2307/40041279 Cited on page(s) 19

[76] F. McSherry and I. Mironov. Differentially private recommender systems: building privacy into the netflix prize contenders. In *ACM SIGKDD International Conference on Knowledge Discovery and Data Mining (KDD)*, 2009. DOI: 10.1145/1557019.1557090 Cited on page(s) 40

[77] G. M. Namata, H. Sharara, and L. Getoor. A survey of link mining tasks for analyzing noisy and incomplete networks. In J. H. Philip S. S. Yu and C. Faloutsos, editors, *Link Mining: Models, Algorithms, and Applications*. Springer, 2010. DOI: 10.1007/978-1-4419-6515-8 Cited on page(s) 17, 18, 20

[78] A. Narayanan and V. Shmatikov. Robust de-anonymization of large sparse datasets. *IEEE Symposium on Security and Privacy*, 2008. DOI: 10.1109/SP.2008.33 Cited on page(s) 13, 38, 40

[79] A. Narayanan and V. Shmatikov. De-anonymizing social networks. In *IEEE Symposium on Security and Privacy*, 2009. DOI: 10.1109/SP.2009.22 Cited on page(s) 12, 34

[80] J. Neville and D. Jensen. Iterative classification in relational data. In *AAAI Workshop on Statistical Relational Learning*, 2000. Cited on

[81] A. S. Ogale and Y. Aloimonos. Shape and the stereo correspondence problem. *International Journal of Computer Vision*, 65:147–162, 2005. DOI: 10.1007/s11263-005-3672-3 Cited on page(s) 17

[82] A. Popescul and L. H. Ungar. Statistical relational learning for link prediction. In *IJCAI Workshop on Learning Statistical Models from Relational Data*, 2003. Cited on page(s) 20

[83] K. Puolamäki, S. Hanhijärvi, and G. C. Garriga. An approximation ratio for biclustering. *Inf. Process. Lett.*, 108(2):45–49, 2008. DOI: 10.1016/j.ipl.2008.03.013 Cited on page(s) 58

[84] M. J. Rattigan and D. Jensen. The case for anomalous link discovery. *SIGKDD Explorations Newsletter*, 7(2):41–47, 2005. DOI: 10.1145/1117454.1117460 Cited on page(s) 20

[85] R. Reeder, L. Bauer, L. Cranor, M. Reiter, K. Bacon, K. How, and H. Strong. Expandable grids for visualizing and authoring computer security policies. In *CHI*, 2008. DOI: 10.1145/1357054.1357285 Cited on page(s) 59

[86] J. Rissanen. Modeling by shortest data description. *Automatica*, 14:465–471, 1978. DOI: 10.1016/0005-1098(78)90005-5 Cited on page(s) 58

[87] B. M. Rubin and E. G. Fitzsimmons. Social-networking sites viewed by admissions officers. *Chicago Tribune*, September 20 2008. Cited on page(s) 1

[88] P. Samarati. Protecting respondents' identities in microdata release. *IEEE Transactions on Knowledge and Data Engineering (TKDE)*, 13(6):1010–1027, 2001. DOI: 10.1109/69.971193 Cited on page(s) 26

[89] P. Samarati and L. Sweeney. Generalizing data to provide anonymity when disclosing information (abstract). In *PODS*, page 188, 1998. DOI: 10.1145/275487.275508 Cited on page(s) 26

[90] P. Sen, G. M. Namata, M. Bilgic, L. Getoor, B. Gallagher, and T. Eliassi-Rad. Collective classification in network data. *AI Magazine*, 29(3):93–106, 2008. DOI: 10.1145/1217299.1217304 Cited on page(s) 18

[91] H. Sharara and L. Getoor. Group detection. *Encyclopedia of Machine Learning*, 2010. Cited on page(s) 20

[92] W. M. Soon, H. T. Ng, and D. C. Y. Lim. A machine learning approach to coreference resolution of noun phrases. In *Association for Computational Linguistics*, 2001. DOI: 10.1162/089120101753342653 Cited on page(s) 17

[93] K. Strater and H. Lipford. Strategies and struggles with privacy in an online social networking community. In *British Computer Society Conference on Human-Computer Interaction*, 2008. Cited on page(s) 44, 59

[94] L. Sweeney. Achieving k-anonymity privacy protection using generalization and suppression. *International Journal of Uncertainty*, 10(5):571–588, 2002. DOI: 10.1142/S021848850200165X Cited on page(s) 13, 26

[95] B. Taskar, P. Abbeel, and D. Koller. Discriminative probabilistic models for relational data. In *Conference on Uncertainty in Artificial Intelligence (UAI)*, 2002. Cited on page(s) 18, 19

[96] S. Trepte and L. Reinecke. *Privacy Online*. Springer, 2011. DOI: 10.1007/978-3-642-21521-6 Cited on page(s) 3

[97] J. Vaidya, C. Clifton, and Y. Zhu. *Privacy Preserving Data Mining*. Springer, 2006. Cited on page(s) 3

[98] N. Vuokko and E. Terzi. Reconstructing randomized social networks. In *SIAM International Conference on Data Mining (SDM)*, 2010. Cited on page(s) 36

[99] G. Wondracek, T. Holz, E. Kirda, and C. Kruegel. A practical attack to de-anonymize social network users. In *IEEE Symposium on Security and Privacy*, 2010. DOI: 10.1109/SP.2010.21 Cited on page(s) 12, 15

[100] L. Wu, X. Ying, and X. Wu. Reconstruction of randomized graph via low rank approximation. In *SIAM International Conference on Data Mining (SDM)*, 2010. Cited on page(s) 36

[101] X. Wu, X. Ying, K. Liu, and L. Chen. A survey of algorithms for privacy-preserving social network analysis. In C. Aggarwal and H. Wang, editors, *Managing and Mining Graph Data*. Kluwer Academic Publishers, 2009. Cited on page(s) 3, 28, 35

[102] X. Ying and X. Wu. Randomizing social networks: a spectrum preserving approach. In *SIAM International Conference on Data Mining (SDM)*, 2008. Cited on page(s) 12, 34, 36

[103] T. J. Ypma. Historical development of the Newton-Raphson method. *SIAM Rev.*, 37(4):531–551, 1995. DOI: 10.1137/1037125 Cited on page(s) 55

[104] P. S. Yu, C. Faloutsos, and J. Han, editors. *Link Mining: Models, Algorithms and Applications*. Springer-Verlag, 2010. DOI: 10.1007/978-1-4419-6515-8 Cited on page(s) 17

[105] E. Zheleva and L. Getoor. Preserving the privacy of sensitive relationships in graph data. *KDD Workshop on Privacy, Security, and Trust in KDD (PinKDD)*, 2007. DOI: 10.1007/978-3-540-78478-4_9 Cited on page(s) 15, 33, 36, 37

[106] E. Zheleva and L. Getoor. To join or not to join: the illusion of privacy in social networks with mixed public and private user

profiles. In *International World Wide Web Conference (WWW)*, 2009. DOI: 10.1145/1526709.1526781 Cited on page(s) 13, 41

[107] B. Zhou and J. Pei. Preserving privacy in social networks against neighborhood attacks. In *IEEE International Conference on Data Engineering (ICDE)*, 2008. DOI: 10.1109/ICDE.2008.4497459 Cited on page(s) 12, 28, 33, 34, 35, 37

[108] B. Zhou, J. Pei, and W.-S. Luk. A brief survey on anonymization techniques for privacy preserving publishing of social network data. *SIGKDD Explorations Newsletter*, 10(2), 2009. DOI: 10.1145/1540276.1540279 Cited on page(s) 3

[109] L. Zou, L. Chen, and M. T. Özsu. K-automorphism: A general framework for privacy preserving network publication. In *International Conference on Very Large Databases (VLDB)*, 2008. Cited on page(s) 12, 28, 33, 34, 35, 42

[110] M. Zuckerberg. An Open Letter from Facebook Founder˙ Mark Zuckerberg. Facebook Blog, December 1 2009. Cited on page(s) 44

Authors' Biographies

ELENA ZHELEVA

Elena Zheleva is a Data Scientist at LivingSocial. She received a Ph.D. in Computer Science from the University of Maryland, College Park in 2011. Her research interests lie in data mining and machine learning for social networks and social media, focusing on statistical models for prediction, evolution, and privacy. She has served on the Program Committees for KDD, AAAI, and CIKM.

EVIMARIA TERZI

Evimaria Terzi is an Assistant Professor in the Department of Computer Science at Boston University. She received a Ph.D. in Computer Science from the University of Helsinki in 2007 and an M.S. from Purdue University in 2002. Before joining Boston University in 2009, she was a Research Scientist at IBM Research. Her work focuses on algorithmic data mining, with emphasis on time-series and social-network analysis. Evimaria has received the Microsoft Faculty Fellowship, and has been in the PC and Senior PC of many data-mining and database conferences including KDD, VLDB, and SIGMOD.

LISE GETOOR

Lise Getoor is an Associate Professor in the Computer Science Department at the University of Maryland, College Park. She received her Ph.D. from Stanford University in 2001. Her research interests include machine learning and reasoning under uncertainty; in addition she works in data management, visual analytics, and social network analysis. She is a board member of the International Machine Learning Society, and co-chaired ICML 2011. She has served as associate editor for ACM Transactions of Knowledge Discovery from Data, the Machine Learning Journal, and JAIR, on the AAAI Executive Council, and on the PC or senior PC of conferences including AAAI, ICML, KDD, SIGMOD, UAI, VLDB, and WWW.

Printed in the United States
by Baker & Taylor Publisher Services